Carceral Mobilities

Mobilities research is now centre stage in the social sciences with wide-ranging work that considers the politics underscoring the movements of people and objects, critically examining a world that is ever on the move.

At first glance, the words 'carceral' and 'mobilities' seem to sit uneasily together. This book challenges the assumption that carceral life is characterised by a lack of movement. *Carceral Mobilities* brings together contributions that speak to contemporary debates across carceral studies and mobilities research, offering fresh insights to both areas by identifying and unpicking the manifold mobilities that shape, and are shaped by, carceral regimes. It features four sections that move the reader through the varying typologies of motion underscoring carceral life: tension; circulation; distribution; and transition. Each mobilities-led section seeks to explore the politics encapsulated in specific regimes of carceral movement.

With contributions from leading scholars, and a range of international examples, this book provides an authoritative voice on carceral mobilities from a variety of perspectives, including criminology, sociology, history, cultural theory, human geography, and urban planning. This book offers a first port of call for those examining spaces of detention, asylum, imprisonment, and containment, who are increasingly interested in questions of movement in relation to the management, control, and confinement of populations.

Jennifer Turner is a Lecturer in Human Geography at the University of Liverpool, UK. Her research is concerned with spaces, practices, and representations of incarceration, past and present. Jennifer has published widely in the fields of carceral geography and criminology. She is the author of *The Prison Boundary: Between Society and Carceral Space* (Palgrave Macmillan, 2016).

Kimberley Peters is a Lecturer in Human Geography at the University of Liverpool, UK. Kimberley's research analyses the governance of mobilities at sea. Most recently she has pursued this interest through interrogating the politics of mobilities aboard the prison ship (with Jennifer Turner) and via a study of the formulation of maritime regulatory apparatus (funded by the Leverhulme Trust).

Routledge Studies in Human Geography

This series provides a forum for innovative, vibrant, and critical debate within Human Geography. Titles will reflect the wealth of research which is taking place in this diverse and ever-expanding field. Contributions will be drawn from the main sub-disciplines and from innovative areas of work which have no particular sub-disciplinary allegiances.

For a full list of titles in this series, please visit www.routledge.com/series/SE0514

Carceral Mobilities

Interrogating Movement in Incarceration

**Edited by Jennifer Turner and
Kimberley Peters**

Routledge
Taylor & Francis Group

LONDON AND NEW YORK

First published 2017
by Routledge

2 Park Square, Milton Park, Abingdon, Oxfordshire OX14 4RN
711 Third Avenue, New York, NY 10017

Routledge is an imprint of the Taylor & Francis Group, an informa business

First issued in paperback 2018

British Library Cataloguing-in-Publication Data
A catalogue record for this book is available from the British Library

Library of Congress Cataloging in Publication Data
A catalog record for this book has been requested

ISBN: 978-1-138-18404-6 (hbk)
ISBN: 978-1-138-38490-3 (pbk)

Typeset in Times New Roman
by diacriTech, Chennai

For Aberystwyth,
where this journey began

Contents

PART IV
Transition 191

Figures

Tables

Contributors

Roberta Altin is Assistant Professor of Cultural Anthropology at the Department of Humanistic Studies, University of Trieste. Her research has mainly focused on transnational migrations, museum ethnography, and media anthropology. Her most recent books include *Destini incrociati. Migrazioni tra località e mobilità* (with A. Guaran, F. Virgilio, 2014); *Coricama, lo specchio della comunità* (2012); *Ordinarie migrazioni* (ed. with F. Virgilio, 2011); *L'intervista con la telecamera: giornalismo, documentario e ricerca socio-antropologica* (ed. with P. Parmeggiani, 2008); and *Nuove frontiere della rappresentazione digitale* (ed. with P. Parmeggiani, 2007).

Elizabeth Brown is an Associate Professor in the Department of Criminal Justice Studies at San Francisco State University and Director of the School of Public Affairs and Civic Engagement. She holds a doctorate in Geography with a certificate in Law and Society Studies and a Master's degree in Cultural Studies. Her research examines the intersections between urban neighbourhoods and crime control policies, and explores how depictions of crime and disorder influence juvenile justice policy, life histories of urban youth, and socioeconomic and racial inequality in the US.

Kate Coddington received her PhD in Geography from Syracuse University in 2014 and is a Lecturer in Human Geography at Durham University. Her ethnographic research in the Indian Ocean region focuses on public policy dealing with migrants and postcolonial governance that influence processes of geopolitics, bordering, and citizenship. This has been published in journals, including *Social and Cultural Geography* and *Progress in Human Geography*.

Deirdre Conlon is Lecturer in Critical Human Geography at the University of Leeds. Her research examines tensions around immigration enforcement/management practices and immigrants' everyday experiences. Current research and writing focus on the internal infrastructure and dynamics that gird immigration detention (in collaboration with Nancy Hiemstra) and the migrant as object of ontological insecurity. In addition to recent articles in *Annals of the Association* *of American Geographers* and *Citizenship Studies*, she is co-editor of the volume *Carceral Spaces: Mobility and Agency in Imprisonment and Detention* (Ashgate, 2013).

Elaine Fishwick is a Sydney-based freelance researcher and writer with many years of academic experience teaching social policy, criminology, and human rights as well as working in policy with governments and non-government organisations in the field of social and legal justice. Elaine is a co-author (with Leanne Weber and Marinella Marmo) of the book *Crime, Justice and Human Rights* (Palgrave, 2014) and co-editor of the recent *Routledge International Handbook of Criminology and Human Rights* (2016).

James Gacek is a doctoral candidate in the School of Law at the University of Edinburgh. Situated within broader research interests in prison sociology, critical criminology, and carceral geography, his PhD research focuses upon male experiences of imprisonment within Manitoba Corrections, community reintegration, and the territorial stigmatization of marginalized neighbourhoods in Canada. James is an American Sociological Association Student Paper Award winner (2014).

Nick Gill is a Professor of Human Geography at Exeter University. He was written on punitive mobilities, asylum, and migration issues. His books include *Mobilities and Forced Migration* (Routledge, 2014), *Carceral Spaces* (Ashgate, 2013), and *Nothing Personal? Geographies of Governing and Activism in the British Asylum System* (Wiley-Blackwell, 2016).

Lirio Gutiérrez Rivera is an Assistant Professor at the Department of Political Science, Universidad Nacional de Colombia (National University of Colombia). Her research includes urban violence, contemporary prisons, youth gangs, social mobility, and state responses to crime and violence in Latin America, particularly Honduras and Colombia. Her current work focuses on urban politics and gender in Medellin, Colombia. Her monograph, *Territories of Violence*, was published in 2013 with Palgrave.

Kirsty Greenwood is a PhD candidate in Criminology and Graduate Teaching Assistant at Liverpool John Moores University. She is also a Visiting Lecturer in Criminology and Sociology at Chester University. Kirsty's Masters thesis explored the regulation of 'deviant' women within the semi-penal institution of Liverpool Female Penitentiary (1809–1921). Her research focused upon the coerced physical and social mobilisation underpinning admissions into Liverpool Female Penitentiary and the subsequent internal carceral regimes of feminisation, domestication, infantilisation, and Christianisation. Kirsty's wider research interests include gender-specific community punishment, gendered penal regimes, and enforced social mobilities.

Nancy Hiemstra is Assistant Professor in the Department of Women's, Gender, and Sexuality Studies at Stony Brook University in Stony Brook, New York. Her research focuses on the sociocultural and political consequences of human mobility, migration policy-making, and embodied experiences of (in)security. Current projects and publishing examine the geopolitical reverberations of US immigration enforcement in Latin America and (in collaboration with Deirdre Conlon) the micro-economies of immigration detention facilities in the United States. With Deirdre Conlon, she is co-editor of the volume *Intimate Economies of Immigration Detention: Critical Perspectives* (Routledge, 2016).

Avril Maddrell is Associate Professor in Social and Cultural Geography at the University of the West of England, Bristol, and Co-Editor of the journal *Gender, Place and Culture*. Her current research interests include spaces and practices of mourning and remembrance and sacred mobilities. Key publications include co-edited volumes: *Sacred Mobilities* (Ashgate, 2015) and *Deathscapes: Spaces for Death, Dying, Mourning and Remembrance* (Ashgate, 2010) as well as monographs on *Charity Shops: Retailing, Consumption and Society* (Routledge, 2002, with Sue Horne) and *Complex Locations: Women's Geographical Work in the UK 1850–1970* (RGS-IBG/Wiley-Blackwell, 2009).

Emma Marshall is a PhD Researcher in Human Geography at the University of Exeter, and is a member of the Spatial Responsibilities Research Group, the South West Doctoral Training Centre, and the Foundation for European Progressive Studies Young Academics Network. Her research draws on her professional experience of working with refugee communities, and is funded by the ESRC. Emma studied Politics and International Studies at the University of Warwick, before completing an LLM in International Human Rights Law and an MRes in Critical Human Geographies at the University of Exeter.

Katie Maher studies and teaches across disciplines in the areas of critical race and whiteness studies, Aboriginal studies, and justice studies. She is currently completing a PhD at the University of South Australia, where her thesis provides a critical analysis of how race has operated across the Australian railways. She teaches Aboriginal Studies at Flinders University, Adelaide.

Peter Merriman is a Professor of Geography in the Department of Geography & Earth Sciences at Aberystwyth University. He is a leading scholar on the geographies of mobilities, and he has authored or edited five books, including *Mobility, Space and Culture* (Routledge, 2012) and *The Routledge Handbook of Mobilities*

(2014, co-edited). He is an Associate Editor of *Transfers: Interdisciplinary Journal of Mobility Studies*, and a member of the editorial boards of *Mobilities* and *Applied Mobilities*.

Bénédicte Michalon is a Senior Research Fellow at the CNRS (Centre National de la Recherche Scientifique), Bordeaux, France. Having previously worked on ethnic migrations, transborder petty trade, and seasonal labour foreign force in agriculture, her current research is concerned with migrations and governmentality, with fieldwork in Romania and France. She focuses on detention and house arrest for foreigners awaiting deportation, asylum seekers' housing in rural areas, and the everyday practices of migrants control. Her ANR and Aquitaine Regional Council research project *TerrFerme* was one of the first French-speaking interdisciplinary teams into spatial readings of confinement, based on crossed investigations on prison, detention, and labour camps (http://terrferme.hypotheses.org/).

Claudio Minca is Professor and Head of the Cultural Geography Department at Wageningen University. He is also Visiting Professor at the University of London, the University of Colorado in Boulder, and Rikkyo University, Tokyo. His current research centres on three major themes: tourism and travel theories of modernity; the spatialisation of (bio)politics; and the relationship between modern knowledge, space, and landscape in postcolonial geography. His most recent books are *Hitler's Geographies* (with P. Giaccaria, 2016), *On Schmitt and Space* (with R. Rowan, 2015), *Moroccan Dreams* (with L. Wagner, 2015), and *Real Tourism* (with T. Oakes, 2011).

Christophe Mincke is the Director of the Criminology Department of the (Belgian) National Institute of Criminalistics and Criminology and a Professor at the Université Saint-Louis of Brussels. His work has focused on subjects that include the public prosecutor and alternative procedures for conflict resolution. He is presently developing a discussion on mobility as a paradigm for current social mutations.

Dominique Moran is Reader in Carceral Geography at the University of Birmingham, UK. She currently holds over £1m of ESRC funding for research into the design and experience of carceral spaces, and she is author of *Carceral Geography: Spaces and Practices of Incarceration* (Ashgate, 2015) and an editor of *Carceral Spaces: Mobility and Agency in Imprisonment and Migrant Detention* (Ashgate, 2013) and *Historical Geographies of Prisons* (Routledge, 2015).

Patrycja Pinkowska is a PhD student in Human Geography at the University of Exeter. Her ESRC-funded research with deportable migrants examines the experiences of and responses to punitive migration policies in the UK. Patrycja holds an

MSc in Gender and International Relations. Before embarking on a PhD project, she spent five years working with organisations supporting asylum seekers and refugees.

Anna Schliehe is a post-doctoral researcher in Criminology at the University of Cambridge and undertook her PhD project in Human Geography at the University of Glasgow. Her research on the Scottish criminal justice system and its responses to young women in particular is informed by both carceral geography and criminological scholarship. Anna is interested in understanding the nature and experience of closed spaces, connecting empirical to conceptually challenging research.

Alex Tepperman is a doctoral candidate in the University of Florida's Department of History, where he works on the social history of American prison communities. He has published widely in this area of criminal and penal history, including the co-authoring the 2013 textbook *Deviance, Crime, and Control: Beyond The Straight and Narrow* (3rd edition, Oxford University Press). He is a graduate and former Junior Visiting Fellow of the University of Toronto Center for Criminology and Sociolegal Studies.

Joaquín Villanueva is an Assistant Professor in the Department of Geography at Gustavus Adolphus College in Minnesota. He obtained his PhD in Geography at Syracuse University and has since expanded his research interests to the fields of political, urban, carceral, and legal geographies. Through a deeper engagement with the right to the city, critical citizenship, and geographies of security and policing scholarships, Joaquín has investigated the spatiality of the criminal justice system in the French *banlieue*.

Michael Wearing is a highly experienced social work educator, social researcher, social policy analyst, and teaching and research manager. He has published six books and over 50 book chapters and journal articles. Michael initially practiced as a social worker and community worker in the area of intellectual disability prior to his academic career. As a Senior Lecturer at the University of New South Wales, he has direct experience and knowledge of academic research and in developing and evaluating programs for social policy, social work, youth work, and juvenile justice and welfare practice.

Foreword

Dominique Moran

We were brought up to the train. It was a freight train for transporting ore. The cars were simply giant wooden boxes for coal, with a narrow ladder on the side. But this ladder was mounted at head level... The guards went crazy. Four of them stood at the door of the car, pushing and beating us, 'Faster, faster!'... The car was filling very fast. At first we did not realise how awfully overcrowded this car was. And we would have to travel in this car for almost a month.

There were no toilets there, no bucket..., there was nowhere to wash. ... the stench and filth were horrendous. The lice, which appeared almost instantaneously, were horrible. There were so many of them, of these lice, that when everybody in the car went to sleep, it seemed that the people were still moving, boiling like water in a pot.

There was a partition in the car. The very first day, they put the dying behind this partition. If opening the doors and taking out dead bodies was a rarity at the beginning, by the sixth or seventh day it was a daily occurrence. The guards would...throw some corpses out through the...gates which are normally opened to unload coal. Usually these corpses were dragged a short distance off to the side so that the train would not catch and drag them.

– Roza Zelmanovna Vetukhnovskaia (1904–1993) (reproduced in Shapovalov, 2001: 159–160)

Accounts like Roza Zelmanovna's, which bring viscerally to life the almost unspeakable conditions of prison transportation in the wartime USSR, remind us that incarceration is a practice and an experience that is neither spatially bounded, nor spatially fixed. The confinement, discipline, control, degradation, maltreatment, and neglect which can characterise incarceration do not require immobilisation for their effects to be felt, and it is often the case that the forms of movement inherent in carceral practices themselves accentuate the 'pains' of imprisonment (Sykes, 1958).

Carceral geography's emergence as a sub-discipline was heavily influenced by the 'new mobilities paradigm' (Sheller and Urry, 2006), such that the recognition that 'contemporary practices of imprisonment are characterized by [the] tensions between apparent fixity and forced mobility' (Martin and Mitchelson, 2009: 461) has been almost a mainstay of geographical research into carceral environments since its inception. That said, this collection of work explicitly interrogating *Carceral Mobilities* puts down a marker about the ways in which carceral geographers view the spaces and practices of confinement. And at the same time, it both illustrates the richness of interpretation and analysis which is enabled by considering the mobilities inherent in the carceral, and speaks back to mobilities scholarship from a research context whose exploration has already challenged the seemingly innate connection drawn between mobility, rights, and autonomy (Gill, 2009; Moran et al., 2012).

The astonishing breadth of this collection, in terms of the times and spaces encompassed by its constituent chapters, is matched only by the richness of the language used to designate both the mobile and the carceral. We read of practices of leaving, returning, immobilisation, exit, entry, enclosure, confinement, mobilisation, moving, displacement, propelling, travelling, circulating, removal, navigation, drift, passing through, riding, departing, mooring, manoeuvring, release, deporting, and departure, which take form within, between, into, out of, away from, beyond, outwith, and across spaces and circumstances termed carceral, semi-carceral, free, in-between, segregated, under curfew, constraining, curtailed, controlled, disciplined, and managed. The contributing authors pick their terms carefully, each negotiating the intricate, manifold relationship between confinement and mobility which, in essence, drives the project of this book.

Amongst the successes of this collection is the way in which it entreats us to look more closely and attentively not just at the carceral-as-mobile, and the mobile-as-carceral, but at the multi-scalar mobilities in which the carceral is implicated and through which it is expressed. Although Roza Zelmanovna's account of the 'boiling' of lice, the violent urgings of guards, the substitution of humans for coal or ore, the casual dragging of corpses, and the squalid degradation of humanity through a notorious prison transportation system supporting Soviet power seem to speak to us from another age, this book reminds us that these tensions, circulations, distributions, and transitions are as critical now as they were then.

University of Birmingham
May 2016

References

Gill N (2009) Governmental mobility: The power effects of the movement of detained asylum seekers around Britain's detention estate. *Political Geography* 28(3): 186–196.

Martin LL and Mitchelson ML (2009) Geographies of detention and imprisonment: Interrogating spatial practices of confinement, discipline, law, and state power. *Geography Compass* 3(1): 459–477.

Moran D, Piacentini L and Pallot J (2012) Disciplined mobility and carceral geography: Prisoner transport in Russia. *Transactions of the Institute of British Geography* 37(3): 446–460.

Shapovalov V (2001) (ed) *Remembering the Darkness: Women in Soviet Prisons*. Oxford: Rowman and Littlefield.

Sheller M and Urry J (2006) The new mobilities paradigm. *Environment and Planning A* 38: 207–226.

Sykes G (1958) *The Society of Captives: A Study of a Maximum-Security Prison*. Princeton, NJ: Princeton University Press.

Acknowledgements

This project emerges from the coming-together of two distinct research interests: one centred on the spatialities of carcerality, the other on the geographies of mobilities. During a summer of discussion, coffee breaks, and a foray into the thought-provoking literature on carceral mobilities highlighted in this introduction, the idea for this book developed. This was in situ with the beginnings of our own project focused on the micro-mobilities of carceral experience, explored through the historical case study of the prison ship. In light of these interests, the aim of this text was to provide a sustained, topically varied, and contextually wide-ranging interrogation of motion in processes and practices of incarceration.

To this end, we are grateful to those who inspired our interests in carceral space, with special thanks to Dominique Moran, who provides the foreword to this collection. We are also grateful to those who sparked our interests in mobilities, with particular thanks to Peter Merriman, who provides the afterword. We would also like to acknowledge John Urry, whose seminal work on mobilities has influenced us both. Projects such as this have been possible only because of his foregrounding, agenda-setting work and the spirit for interdisciplinarity that he fostered.

Indeed, this collection brings together a wide range of scholars – from early career academics to established professors – from the disciplines of geography, criminology, social policy, and history, and we are grateful to these authors for their enthusiasm in the project: the fascinating, challenging, eye-opening chapters they have contributed and the hidden mobilities of carceral worlds which have been revealed. We would also like to extend our thanks to the external reviewers of the collection for their support of the project and their helpful feedback; and to Faye Leerink, Priscilla Corbett, and the rest of the team at Routledge for their assistance in the publication of this book.

Finally, Jen would like to thank Kim for initiating her into the complex world of mobilities with generously provided expertise and gin and tonic in equal measure. Without this wide-ranging knowledge of the nuances of this area of study, the collection would have been unable to challenge the boundaries of conventional carceral studies so effectively. Kim would like to thank Jen for her conscientious management of the project, for her careful editing, and for introducing her to the sub-field of carceral studies and the inspiration this has provided for thinking anew, geographically.

Leicester, 2016

1 Carceral mobilities

A manifesto for mobilities, an agenda for carceral studies

Kimberley Peters and Jennifer Turner

Introduction

Mobilities research is now an established field of academic enquiry (Faulconbridge and Hui, 2016). In the decade since the publication of two seminal papers that positioned mobilities as 'centre stage' in social science research agendas (Sheller and Urry, 2006: 208; Hannam et al., 2006), a proliferation of wide-ranging work on the politics underscoring the movements of people and objects has emerged. As is well documented elsewhere (Adey et al., 2014; Cresswell and Merriman, 2011), this has ranged from a study of technologies of motion (airplanes, trains, buses, cars, and bicycles) to the infrastructures that enable/disable mobility (roads, rails, airports, data centres); the subjects made mobile or immobile by regimes of regulation and control (including commuters, tourists, migrants, military personnel, and so on); and the materialities that shape and are shaped by mobilities (food distribution, fossil fuels, passports, and so forth). What 'mobilities thinking' has come to achieve, therefore, is a critical consideration of a world that is ever 'on the move' (Cresswell, 2006). But with the 'maturity' (Faulconbridge and Hui, 2016: 8) of the 'mobilities paradigm' (Sheller and Urry 2006: 207), how might the study of mobilities *move* forwards? For Faulconbridge and Hui, the future of mobilities research relies on the study of movement – and the politics of movement – remaining 'vibrant, creative and generative' (2016: 1). It relies on a recognition that 'mobilities research is itself on the move'; drawing in new spaces, subjects, events, occurrences, and temporalities to examine through a mobilities framework (Falconbridge and Hui, 2016: 1). This movement, we argue, has motioned scholars towards the study of *carceral mobilities*.

At first glance, the words 'carceral' and 'mobilities' seem to sit uneasily together. Consider the brute physicality of a prison wall. Whether stone, brick, edged with barbed wire, or flanked with surveillance, the boundary between the 'inside' and 'outside' of this particular carceral space also marks an assumed boundary between movement and stasis (Philo, 2014). Outside of these sites is a presumed autonomy of movement, a liberty to move freely. Inside, there is an assumed prohibition of movement, an imposed stasis and inability to move at will. Such a distinction is evident in the conceptual understandings generated in respect of other sites of detention, captivity, and holding: the migrant camp, the detention centre, the quarantine island (to name just a few). Unsurprisingly, then, where there has been a shift in recent years to take seriously regimes of imprisonment, detention, temporary-holding, and captivity (see Conlon, 2011; Loyd et al., 2012; Martin and Mitchelson, 2009; Moran, 2015; Morin and Moran, 2015; Moran et al., 2013; Pallot, 2005; Turner, 2013, 2016) the study of mobilities have been traditionally 'overlooked' (Moran et al., 2012: 446; and also Ong et al., 2014; Philo, 2014). Scholars have, by and large, neglected to ask what a study of movement might offer to understanding environments of relative stasis.

To answer such a question we have to look more carefully to carceral boundaries, and to the concept of *relation* (Adey, 2006). Carceral boundaries – such as prison walls or detention centre fences – mark a further distinction between the visible and invisible which cements the dichotomy between mobility and immobility that has come to define knowledge of carceral systems. Outside is a world that is known. Inside is world few of us will see and which is visualised and known only through media depictions and the imagination (Turner, 2016). These often stark, material boundaries hide the inside, creating an *appearance* of immobility within carceral estates vis-à-vis an *appearance* of hyper-mobility beyond them. Indeed, for Peter Adey, it is relation which is crucial to understanding mobilities. For those embracing a mobile ontology (see Cresswell, 2010), the world is always, wholly 'on the move'. Stasis, stillness, fixity are products of a relationship *with* mobility (in other words, we can only know that there is immobility when comparing it to mobility). Mobilities rely on anchor points and moorings, binding the two together (Hannam et al., 2006). Moreover, though, as Adey contends, mobilities held in relation create 'illusions of mobility and immobility' (Adey, 2006: 83). A sleepy village may *appear* to be immobile, unchanging and static. But this is only because of the 'speed' of change and movement elsewhere. The sleepy village is, of course, not fixed. Through the passage of time, small, almost unnoticed changes occur. Place is mobile, only at a different pace. Returning to sites of incarceration, here the 'inside' appears less mobile than the outside. The 'outside' appears more mobile than the inside. Of course sites of incarceration should not only been understood through the equally problematic dichotomy of inside/outside (rather, as this book shows, carceral spaces reach into, beyond, spill over, muddy, and blur any socially and materially constructed boundaries; see also Turner, 2016) but it remains that carceral spaces evoke a visual trickery – an illusion of immobility, where instead spaces of incarceration are often underscored by mobilities. These are mobilities that are simply *different* than those that might 'normally' be associated with our understandings of movement and motion (Adey, 2006: 83).

The *illusion* (see Adey, 2006) of carceral space as fixed space limits the possibilities of engaging with such sites as a means of unlocking new knowledge of mobilities. It has led to a neglect in the study of mobilities in relation to carceral life – and to a 'weakness' in the study of mobilities as the field has shied away from spaces of apparent fixity (Moran et al., 2012: 446). Yet with mobilities research expressing an explicit focus on the dimensions of *power* embedded in mobile life (including, then, of course how power works to immobilise) carceral settings seem to be fertile ground generating fresh insights into questions of how, why, in what ways are people and objects able, unable, and restricted in their movement. Likewise, mobilities can offer carceral scholars a framework for better understanding the operation of the power that works to confine, contain, detain, immobilise and also *make mobile* incarcerated peoples. In other words, the question is not whether studies of mobilities have a place in the field of carceral studies and vice versa, but rather: How might examinations of mobilities in carceral settings, and carceral examples of im/mobility, help enrich both the study of carceral geographies (see Moran, 2015) and the progression of mobilities research into the next decade? What opportunities, in short, can sites of immobilisation and stasis offer for understanding mobilities and mobile experience, practically and theoretically?

In the introduction to this edited collection we therefore present something of a manifesto for where the mobile field of mobilities studies might move next, alongside an agenda for the future of carceral studies. There is now a wide acknowledgement that '[m]obility is … a constant practical concern in the management of penal systems' (Moran et al., 2012: 449) (see also Gill, 2009; Moran et al., 2013; Mincke and Lemonne, 2014; Mountz, 2011; Philo, 2014). As Alison Mountz and colleagues have noted (2013), no regime of incarceration is without movement. Movement predicates and dictates what it is to be imprisoned, detained, or held captive (see Moran et al., 2013; Gill, 2009; Moran et al., 2012; Peters and Turner, 2015). As such, whilst carceral space may not be the most *apparent* lens through which to explore mobilities, mobility is part and parcel of carcerality. This acknowledgement allows us to ask a host of questions pertaining to past and present manifestations of carceral life: How are the movements of persons and objects enabled and restricted within carceral environments such as prisons, detention camps, and asylum centres? Through what means does movement occur between such sites, as people, contraband, and ideas are transported between spaces of confinement? In what ways do technologies of incarceration, legal and regulative apparatus, and economic systems impact who and what can move – and where – in/between carceral spaces? How are identities made mobile within carceral regimes, and in what ways do virtual and imaginative capacities shape new possibilities for movement? Ultimately, what systems of power shape these mobilities and what does this mean for better understanding methods of incarceration, and the politics of mobilities? These are the questions central to this edited collection.

In proposing that studies carceral mobilities might be a means of generating fresh discussion for mobilities scholars and carceral geographers alike, we are not arguing for carceral mobilities to emerge as a tightly bounded subfield under the banner of 'mobilities' but rather as a means of continuing the interdisciplinary project that the mobilities paradigm has so far generated (see Faulconbridge and Hui, 2016). Mobilities has been a project uniquely cross-and transdisciplinary in scope. A study of carceral mobilities is inspired by this lineage and provides a means of extending this trajectory into the future, weaving together scholars working in the cognate fields of criminology, sociology, international relations, and human geography. Drawing on a wide range of perspective and expertise, but sharing – fundamentally – a concern with mobility, this collection seeks to ask how the key tenets of mobilities thinking might be reanalysed in the context of the carceral; and how the carceral can be better understood through an attention to mobilities. Accordingly, this book hopes to offer the potential to radically contribute to studies of mobilities as well as those centred on the politics of incarceration. In what follows we outline this project in greater detail, summarising the contributions this book makes to our understandings of carceral mobilities, before suggesting how such research might extend further in the future.

The chapters

The book to follow attends to the relationship between carcerality (or, we might argue, *carceralities*, as we recognise the conditions, qualities, and experiences of carceral life to be multiple not singular) and mobilities. It does so through 16 carefully selected chapters, authored by geographers, criminologists, legal scholars, sociologists, and practitioners working in (and with) social policy. These chapters each identify and unpick a range of mobilities that shape (and are shaped by) carceral regimes. They speak to contemporary debates across carceral studies and mobilities research, offering fresh insights to both areas of concern. Importantly, the book moves this discussion internationally – from the Global North to the Global South – providing an examination of carceral mobilities that are themselves not singular but which are couched in a variety of specific, yet networked, spatial contexts (including Australia, the United States, Latin America, France, Britain, Romania, and Italy). Moreover, the book is organised in four sections that move the reader through the varying typologies of motion underscoring carceral life: tension, circulation, distribution, and transition (see Peters, 2015; and also Cresswell, 2010). Each mobilities-led section seeks to explore the politics encapsulated in specific, yet fluid, regimes of carceral movement. Accordingly, as the field of carceral studies gains momentum (see notably Moran et al., 2013; Moran, 2015; and Morin and Moran, 2015) and the social sciences continue to analyse a world of movement and mobility (see Adey et al., 2014; Faulconbridge and Hui, 2016), *Carceral Mobilities* offers a text of international, interdisciplinary scope which contributes to these topical, timely areas of concern.

The chapters in Part One consider how carceral movements are driven by, and are laden with **tensions** – a strain produced between two or more subjects and/or objects. Tension, this section shows, is a force (see Cresswell, 2010) that produces particular mobile outcomes. Specifically the authors each consider the tensions between those who are incarcerated and those who seek to incarcerate them. In Chapter Two Kate Coddington explores how, since the Northern Territory Emergency Response legislation of 2007, Aboriginal Australians in this area have experienced a range of interventions targeting their communities. Meanwhile, hundreds of asylum seekers have experienced mandatory, indefinite detention in the same region, earning it the name 'Detention Capital of Australia'. Coddington argues that in spite of the apparent differences in these cases of incarceration, similar mobile, 'carceral logics' underscore these different modes of detention.

Crucially for Coddington, these carceral logics are mobile. Regulatory regimes, she posits, move across space and time, and this has violent consequences for both Aboriginal Australians and asylum seekers.

Also attending to the ways in which violence is manifest through carceral regimes, in Chapter Three Roberta Altin and Claudio Minca investigate the semi-carceral operations of detention/hospitality centres for asylum seekers, focusing on the carceral regimes that govern the movement of those housed in the centres. Drawing on the case study of the Gradisca Hospitality Centre for Asylum Seekers in the North East of Italy close to the Slovenian border, Altin and Minca examine the rising tensions surrounding supposed 'unconstrained' mobilities of those seeking asylum who are permitted into 'buffer zones' local to the camp. Here, asylum seekers and migrants alike are seen to create a 'human excess', stimulating a prejudice amongst those 'hosting' them. Whilst perceived as mobile, such individuals are in fact subject to conditions that create heavily regulated and prescribed mobilities – producing a quasi-carceral experience, for largely non-criminal populations. Whilst there is a tension between perceived 'inside' and 'outside' and between those who 'belong' and 'do not belong', Altin and Minca show how asylum seekers also use the mobilities which constrain them to create space for resistance.

Moving back to Australia, Chapter Four turns to youth justice and the tensions between young people and the authorities who seek to govern them through so-called diversionary practices (such as tags, curfews, probation). Elaine Fishwick and Michael Wearing show how such measures, articulated within a landscape of neocolonialism, are as incarcerating as prison, raising tensions regarding where regimes of carcerality apply (and do not apply). In particular, they demonstrate that diversionary practices for ensuring youth justice within New South Wales hinge on mobilities – the perceived need to govern 'unruly' mobilities through the control of how, where, and when young people can and cannot move. Like Altin and Minca they show how the existence of 'carceral logics' (to borrow from Coddington) creates 'liminal' spaces of semi- or quasi-confinement for those subject to such measures.

To conclude the section, in Chapter Five, Bénédicte Michalon also pays attention to semi-quasi, tension-filled regimes of carcerality via an investigation of the 'continuum' of mobilities (borrowing from Gill, 2009) for migrants and asylum seekers in Romania. Here Michalon shows how regimes of so-called tolerance create specific modes of mobility that allow neither unrestricted movement nor total confinement. 'Tolerance' refers to a very temporary right to remain in Romanian territory for irregular migrants who are neither legal nor who can be deported. 'Tolerance', Michalon reveals, is fraught with tension – allowing irregular migrants the right to move but only under specific, self-regulated conditions.

Following on, Part Two explores considers various forms of **circulation** – physical, material, and imaginative – that operate in carceral space and across the boundary between prison and society (see also Turner, 2016). The authors each explore how carceral life depends upon, is forged and framed by, and is threated via circulations. As such, these authors urge us to rethink how mobility is perceived and encountered in carceral space. Writing from a perspective of prison sociology, in Chapter Six James Gacek contends that much can be learned from the workings of space and motion by understanding inmates' identity constructions and their ability to cope with stress within prison. Using qualitative interviews with ten men who have experienced incarceration in Manitoba, Canada, Gacek draws attention to the psychosocial dimensions of spatiality, the role of inner space, and importance of daydreaming for prisoners to 'escape' and move beyond their physical existence 'inside' prison. Accordingly he explores an imaginative circulation of personhood from inside to outside and back again, through the power of the mind. He suggests that the use of imagination by inmates is crucial for allowing inmates to adapt to – and move beyond – the carceral structure.

Relatedly, in Chapter Seven Alex Tepperman unpacks the circulation of ideas through the spreading of a 'convict code' in early-twentieth-century American prisons. His chapter tracks the spread of an anti-institutional ideology common among American inmates from the 1920s to the mid-twentieth century. In addressing the informal movement of inmate culture through the mass movement of prisoners nationwide, Tepperman uses historiographical methods to explore how such a code and culture was situated amidst dramatic changes to public policy and penal architecture, as well as increasing racial and cultural heterogeneity of state and federal prison communities. Tepperman illustrates how the creation and circulation of the convict code produced meaningful segmental bonding among inmates, which ultimately led to the further mobilisation of the code as it was shared and spread.

Following from this, in Chapter Eight Deirdre Conlon and Nancy Hiemstra further consider the spread of ideas examining how ideologies which link together migrants and criminals – positioning them as 'one and the same' – are mobilised, circulated, and dispersed. In doing so they demonstrate how this circulation of carceral ideologies has profound impacts for producing exclusions in wider society. Tracking and tracking policy and media reports, like so many authors in the collection, they demonstrate how resistant practices can challenge mobilities that constrain and limit individuals and collective groups. Indeed, Conlon and Hiemstra argue that 'counter-mobilisations' are needed to resist circulations of popular perceptions.

Finally, Anna Schliehe considers more concrete circulations in Chapter Nine, unpacking the significance of objects and their circulation between prison and the home. Embracing an object-orientated ontology, Schliehe explores the role of objects in processes of place-making for prison inmates in Scotland. Whilst incarcerated, the ability for inmates to physically move from place to place around the prison estate is often restricted. Where mobility is possible it follows patterns of strict routine, often associated with social hierarchy or status. As Schliehe shows, this pattern of movement and standstill, what is permitted and what is denied, extends to prisoners' possessions. Drawing on findings from qualitative research with young female prisoners in Scotland, Schliehe shows how (drawing on Goffman) objects circulate or 'float' around the prison – passed, hand-to-hand – but notably, also how circulations of personal objects from the home to the prison allow those incarcerated to gain a sense of self amidst regimes that often strip inmates of their individuality.

Dovetailing the section on circulation, the book next turns to questions of **distribution**, exploring in Part Three how carceral mobilities are produced through distributive mechanisms which create a dispersal of populations and regulations – and ultimately power – in/between the prison and society. In Chapter Ten, Emma Marshall, Patrycja Pinkowska, and Nick Gill explore the possibilities of virtual space and social-media technologies for distributing stories of carceral experience, in order to reshape understandings of migrant and asylum seeker experiences. They demonstrate the mobilising capacities of programmes such as Twitter for 'giving voice' to silenced populations, whilst simultaneously drawing out the issues of access to such technologies where the use of phones and computers is often severely limited. Focusing on web-based anti-detention campaigns in the UK, the authors argue that the Internet opens up the possibility for activism to take new forms, creating alternative forms of social and political mobilisation for vulnerable groups.

In Chapter Eleven, Joaquín Villanueva continues to consider distribution as a form or typology of motion, investigating prosecutorial spaces as under-analysed sites in carceral studies. In this chapter, Villanueva recognises that prosecutorial spaces are crucial sites where decisions are made, which constrain the mobility of individuals elsewhere. In other words, it is prosecutorial spaces that distribute regimes of governance that impact mobility. More so, Villanueva shows how prosecutorial spaces are not fixed in space – but are themselves mobile and distributed. Drawing on the example of the Parisian *banlieue* of Seine-Saint-Denis, Villanueva illustrates the existence of prosecutorial mobility – where courts, chief prosecutors, and decisions makers 'pop up' and move flexibly across the urban landscape with the intention of mobilising local and regional resources destined to contain or detain individuals located in 'sensitive' neighbourhoods. In short, Villanueva argues that this form of mobility has become an important tactic for the exercise of judicial authority.

Katie Maher, in Chapter Twelve, turns her attention to the infrastructure of the railroad in colonial and postcolonial Australia and the use of indigenous, criminalised, and incarcerated labour to build and maintain the rail network. Maher demonstrates how power was distributed along the railroad, containing, confining, and immobilising those who ironically produced this technology of mobility. Drawing on rich archive records and newspaper cuttings, Maher pieces together a story of mobilities that has in turn distributed and cemented ideas about race, status, and belonging. In Chapter Thirteen to follow, Gutiérrez Rivera likewise explores the distribution of power around the Latin American prison in Honduras. Here power does not move from the top-down, in hierarchical structures typical of the prison environment. Rather, guards appoint a small group of inmates known as 'rondines' to supervise the rest of the inmates. Whilst still resting on a structure where some inmates may distribute rules, regulations, allowances, and so forth over others, this chapter subverts our understandings of how mobilities work *within* carceral spaces when those mobilities are governed by inmates themselves. Indeed, Gutiérrez Rivera argues that the Honduras prison establishes power relations based on class, which can then control, regulate and distribute prison resources in ways atypical to prisons in the Global North.

To conclude, the final section of the book turns to the mobile condition of **transition**. Part Four considers the mobilities realised as persons, things, and ideas *shift* from one space, situation, setting, to another. The authors of this section each consider the experiences of – and politics entwined with – transitions *within* carceral space, and *between* the life 'inside' and society 'outside'. Kirsty Greenwood begins the section in Chapter Fourteen by making a contribution to an emerging knowledge of historical carceral mobilities (see also Morin and Moran, 2015). Here, she uncovers a hidden narrative related to the regulation of 'deviant' women through the use of semi-penal institutions designed to enable the social mobility of women for their reintegration back into society. This chapter unpicks the biopolitical efforts employed to transform deviant women into 'respectable' females between 1809 and 1921 in Liverpool, UK. Using the example of Liverpool Female Penitentiary (LFP), Greenwood explores how women were first *removed* from the city, before transitioning back into the city, transformed and rehabilitated. Importantly, Greenwood traces the resistances to these restrictive and idealised regimes and the highly gendered nature of such systematic control.

Next, in Chapter Fifteen, Elizabeth Brown traces how mobility has shaped the history and practices of the juvenile court in the United States. In short, she demonstrates the transition of policies through time which have impacted the mobility of young people via modes of control (for example, curfews), the removal of young people from familial homes, and increased immobilisation through regimes of policing and surveillance. Like Villanueva, Brown challenges the assumption that courts are somehow static points from which decisions are made and measures are laid out. Rather she shows how those decisions and measures shape the mobilities of those implicated in their use. Accordingly, Brown problematises the perception of the court as an inherent 'good', 'saving' young people from the ills of urban life, to instead show how progressive directives – adapted over time – perpetuated inequalities and disadvantaged already marginalised and disenfranchised populations.

In Chapter Sixteen to follow, Avril Maddrell utilises examples from a longitudinal study of a scheme where prisoners on day-release licence 'do time' in the form of Community Service in charity shops. Drawing on in-depth interviews with charity shop managers, volunteers, prisoners, and prison officers, from a male prison in the UK, Maddrell posits the charity shop as a dynamic permeable carceral space, which assists in the transition of licensed prisoners from life in prison, to life back in wider society. Maddrell demonstrates how placements in charity shops contribute to our understanding of the dynamic assemblage of permeable spaces and boundaries that can constitute carceral spaces and practices in the UK and the transitional mobilities associated with the temporal life cycle of a prison sentence.

Finally, Christophe Mincke concludes the collection by developing the concept of a 'mobilitarian ideology' (see also Mincke and Lemmone, 2014). Here, Mincke presents the need for a wholesale transition in how we think of systems of incarceration. In his chapter, Mincke illustrates the ways in which punitive immobility – a fundamental premise and classical aim of regimes of incarceration – is now shifted on its head, where instead, he argues, mobility is the bedrock of carceral systems. Where immobility has been a concept used to legitimise the function of the prison, now it is mobility that legitimises the existence of penal estates in the twenty-first century. Drawing on the Belgian Prison Act of 2005, Mincke shows how – instead of undermining the prison – a mobilitarian ideology is used as a new ground for its renewed and reversed legitimation where the prison is now presented as a mobility-compatible and, even, mobility-based institution.

Moving forwards

With their empirical variety, theoretical diversity, but collective attention to mobilities, the chapters that make up this collection offer mobilities studies an opportunity to assess the ways in which movements are (re)made and (re)produced in sites that have established, concretised, and enduring appearances as ones of *relative* stasis. In turn, the chapters offer carceral studies the potential to unlock further the ways in which mobility is 'centre stage' (to borrow from Sheller and Urry) in regimes that are foregrounded upon ideologies and mechanisms designed to limit and contain. Such a project, this book demonstrates, is worthwhile. Through the chapters to follow we are able to develop a richer understanding of the *unequal mobilities* that pervade society – both today, and in the past (a key objective for mobilities studies, according to Faulconbridge and Hui, 2016). In addition, we can better engage with *mobile futures* (see also Faulconbridge and Hui, 2016), where the future is one that is deniably 'hyper-carceral', with the proliferation of prisons, offshore detention centres, migrant camps and so on, as well as the extension of im/mobilising carceral regimes into everyday spaces and places.

However, we would also posit that there is further work to do to expand the contributions of this book and to enhance our understandings of carceral mobilities. These lines of enquiry are sparked by this very book and the provocations it has generated. Further questions remain to be asked, and answered, concerning the spaces, scales, subjects, objects, methods, and theories that are part and parcel of carceral and mobile life. In terms of spaces, we might ask if there are still sites, contexts, and places where carceral mobilities remain underexplored; for example, the prisoner of war camp, which has been thus far marginalised in the subfield of carceral geography; or rehabilitation centres, where regimes of 'carcerality' might be self-imposed. We might also attend to a host of everyday spaces (which this book points towards) – the street, the home, the workplace – where carceral conditions and qualities emerge through im/mobilising practices. Applying a mobilities framework to an extended collection of carceral spaces might then help us to better think through a question that is still plaguing carceral geographers – *what is the carceral?* (Moran, 2016; Turner and Moran, 2015).

Attention to the spaces of carceral mobilities also requires an attention to scale, a key geographic trope for understanding the operation and dissemination of power. This text has drawn attention to the scales across which carceral mobilities operate – from national-level regimes that work to confine and immobilise (as seen in the example of the railroad construction in Maher's chapter) to the local scale of movements within carceral spaces (typified by internal regimes of motion illustrated in the contribution by Gutiérrez Rivera). The book also draws attention to the linkages of scales, as mobilities within bordered spaces, such as the prison (for example, the spread of a code by word of mouth in a single institution) is circulated nationwide, encompassing many prisons. This example might also be interpreted not via a scalar reading, but rather through a flat ontology (Marston et al., 2005) where carceral mobilities might be understood as networked across space. To comprehend how mobilities and carceralities intersect and form and forge one another, there is a need to pay attention to scales and networks more seriously. On the one hand – as we have suggested elsewhere (see Peters and Turner, 2015) – this requires us to look to the micro-mobilities (see Bissell, 2010) that shape carceral experiences (for example, the internal motions of the body produced through sickness, the rubbing of restraints against skin, and so forth). There is also a need to look at how global politics is shaping regimes of movement which incarcerate – through 'buffer zones' in extra-territorial spaces that hold migrants liminally offshore (see Mountz and Loyd, 2014) to exceptional spaces where terrorists are held, given the suspension and ambiguity of national and international regulations (see Agamben, 2005).

We might also give further thought to the im/mobile subjects that are bound up with carceral regimes. This book presents a range examples: the mobilities and moorings of young people; regimes that racially discriminate; the mobilities of those held in temporary 'hospitality' camps; those imprisoned; asylum seekers; migrants; and also those whom wider society hopes to 'reform'. Whilst these chapters do important work in bringing marginalised and vulnerable groups to the forefront of discussion, they still, inadvertently, speak *for* such groups from the top-down perspective of the researcher. More could be done to give voice to the range of persons and groups who are imbricated in regimes that curtail or limit movement. Other individuals remain absent, and our attention to aged, gendered, and sexed carceral mobilities could be explored further (see Rosenberg and Oswin in respect of the latter, 2015). Moreover – whilst much work focuses on the disciplined and coerced mobilities of those incarceration, more could be done to shed light on the mobilities of those *doing* the incarcerating: prison guards, asylum centre workers, transport services which move prisoners, and so on. This book touches on this through examples of the mobilities central to judicial and prosecutorial roles (see Brown and Villanueva) but there remains space to better understand carceral mobilities from a fuller range of subject-centred perspectives.

Furthermore, whilst this book has pointed towards the politics of mobility in relation to objects (which are often regulated and restricted in carceral spaces, see Schliehe), much more attention to movements of objects in carceral spaces can assist in understanding regimes of security and surveillance in the prison (shedding light on contraband, and its ability to slip past and through technologies designed to prevent boundary penetration). How are drones, for example, utilised to infiltrate the prison and deliver restricted objects? Moreover, how are mobilities within carceral spaces such as the prison shaped by competition over material resources? And how does access to 'things' and 'stuff' depend on evidence of appropriate disciplined mobilities which in turn earn specific rights and privileges? There is also a greater need to focus on 'big' things (see Jacobs, 2006) and consider the materiality of architecture for shaping carceral regimes (see Moran et al., 2016). The role of architecture as part and parcel of carceral mobilities is oddly absent in this collection. Following from this, we might also question how movement is permitted through virtual means for those who are incarcerated (as Marshall et al. describe in this volume). As the example of 'Prison Cloud' – a secure Internet platform for inmates, currently being trialled in Europe – demonstrates, there are might be novel ways of investigating how movement is made possible in carceral environments, as well as highlighting the tensions in such allowances for those who have had everyday liberties legally curtailed.

Finally, we might consider further the methods and theories used to conduct research into carceral mobilities. At the start of this chapter we noted that the terms 'carceral' and 'mobilities' appear to sit in opposition to each other. Although carceral spaces present only an illusion of stasis, it remains a brute fact – as this collection shows – that carceral spaces are filled with restrictions and limitations, curtailments and exceptions. These same constraints impact the access of researchers to carceral spaces in this first instance (see Altin and Minca's chapter). This is an issue not only associated with studies of carceral movement; rather, it is one that impacts the study of carceral space per se. Whilst there are spatial differences in access (demonstrated in this book, where access to prisons in the Global North rely on different permissions from those in the Global South, for example) researchers must themselves abide to rules and regimes that govern their own movement once within such sites. Such limits are, in many cases, necessary to protect vulnerable people, and researchers. Yet the capacity of research to provide understanding, give voice to unheard populations, and to even evoke change demonstrates the need for continued 'carceral' work. We might ask, therefore, how might we actually research carceral mobilities more effectively. Is there a place for mobile methods in researching the mobilities of carceral populations and places? (See Büscher and Urry, 2009.) And following on, what theories might we employ to make sense of such movements in carceral space? Returning again to the start of this chapter, we noted that carceral mobilities offers another opportunity to extend the interdisciplinary project of mobilities studies – and arguably – carceral studies. A study of *carceral* mobilities may provide the opportunity to use tools of analysis from disciplines ranging from criminology to legal studies, which bring new frameworks of understandings for making sense of im/mobilities in a host of carceral settings. In turn, a mobile ontology may allow those working in a range of fields from social policy to history to enhance their examinations of carceral processes and practices. We therefore close this introduction by opening up lines of enquiry that build from this book, which carceral scholars, mobilities scholars, geographers, criminologists, sociologists, and historians may wish to address *together*.

References

Adey P (2006) If mobility is everything then it is nothing: Towards a relational politics of (im)mobilities. *Mobilities* 1(1): 75–94.

Adey P, Bissell D, Hannam K, Merriman P and Sheller M (eds) (2014) *The Routledge Handbook of Mobilities*. Abingdon: Routledge.

Agamben G (2005) *State of Exception*. Chicago, IL: University of Chicago Press.

Bissell D (2010) Vibrating materialities: Mobility–body–technology relations. *Area* 42(4): 479–486.

Büscher M and Urry J (2009) Mobile methods and the empirical. *European Journal of Social Theory* 12(1): 99–116.

Conlon D (2011) Waiting: Feminist perspectives on the spacings/timings of migrant (im) mobility. *Gender, Place and Culture* 18(3): 353–360.

Cresswell T (2006) *On the Move: Mobility in the Modern Western World*. Abingdon: Routledge.

Cresswell T (2010) Towards a politics of mobility. *Environment and planning D: Society and Space* 28(1): 17–31.

Cresswell T and Merriman P (2011) *Geographies of Mobilities: Practices, Spaces, Subjects*. Farnham: Ashgate.

Faulconbridge J and Hui A (2016) Traces of a mobile field: Ten years of mobilities research. *Mobilities* 11(1): 1–14.

Gill N (2009) Governmental mobility: The power effects of the movement of detained asylum seekers around Britain's detention estate. *Political Geography* 28(3): 186–196.

Hannam K, Sheller M and Urry J (2006) Editorial: Mobilities, immobilities and moorings. *Mobilities* 1(1): 1–22.

Jacobs JM (2006). A geography of big things. *Cultural Geographies* 13(1): 1–27.

Loyd JM, Mitchelson ML and Burridge A (2012) *Beyond Walls and Cages: Prisons, Borders, and Global Crisis*. Athens, GA: University of Georgia Press.

Marston AA, Jones JP and Woodward K (2005) Human geography without scale. *Transactions of the Institute of British Geographers* 30(4): 416–432.

Martin LL and Mitchelson ML (2009) Geographies of detention and imprisonment: Interrogating spatial practices of confinement, discipline, law, and state power. *Geography Compass* 3(1): 459–477.

Mincke C and Lemonne A (2014) Prison and (im)mobility. What about Foucault? *Mobilities* 9(4): 528–549.

Moran D (2015) *Carceral Geography: Spaces and Practices of Incarceration*. Farnham: Ashgate.

Moran D (2016) Conceptualising the carceral in carceral geography. Paper presented at the Annual Meeting of the American Association of Geographers, San Francisco, CA.

Moran D, Gill N and Conlon D (eds) (2013) *Carceral Spaces: Mobility and Agency in Imprisonment and Migrant Detention*. Farnham: Ashgate.

Moran D, Piacentini L and Pallot J (2012) Disciplined mobility and carceral geography: Prisoner transport in Russia. *Transactions of the Institute of British Geographers* 37(3): 446–460.

Moran D, Turner J and Jewkes Y (2016) Becoming big things: Building events and the architectural geographies of incarceration in England and Wales. *Transactions of the Institute of British Geographers*. 41(4): 416–428.

Morin KM and Moran D (eds) (2015) *Historical Geographies of Prisons: Unlocking the Usable Carceral Past*. Abingdon: Routledge.

Mountz A (2011) Where asylum-seekers wait: Feminist counter-topographies of sites between states. *Gender Place and Culture* 18(3): 381–399.

Mountz A, Coddington K, Catania RT and Lloyd JM (2013) Conceptualizing detention. Mobility, containment, bordering, and exclusion. *Progress in Human Geography* 37(4): 522–541.

Mountz A and Loyd J (2014) Transnational productions of remoteness: Building onshore and offshore carceral regimes across borders. *Geographica Helvetica* 69(5): 389–398.

Ong CE, Minca C and Felder M (2014) Disciplined mobility and the emotional subject in Royal Dutch Lloyd's early twentieth century passenger shipping network. *Antipode* 46(5): 1323–1345.

Rosenberg R and Oswin N (2015) Trans embodiment in carceral space: Hypermasculinity and the US prison industrial complex. *Gender, Place and Culture* 22(9): 1269–1286.

Pallot J (2005) Russia's penal peripheries: Space, place and penalty in Soviet and post-Soviet Russia. *Transactions of the Institute of British Geographers* 30(1): 98–112.

Peters K (2015) Drifting: Towards mobilities at sea. *Transactions of the Institute of British Geographers* 40(2): 262–272.

Peters K and Turner J (2015) Between crime and colony: Interrogating (im)mobilities aboard the convict ship. *Social and Cultural Geography* 16(7): 844–862.

Philo C (2014) 'One must eliminate the effects of … diffuse circulation [and] their unstable and dangerous coagulation': Foucault and beyond the stopping of mobilities. *Mobilities* 9(4): 493–511.

Sheller M and Urry J (2006) The new mobilities paradigm. *Environment and Planning A* 38(2): 207–226.

Turner J (2013) Disciplinary engagements with prisons, prisoners and the penal system. *Geography Compass* 7(1): 35–45.

Turner J (2016) *The Prison Boundary: Between Society and Carceral Space.* London: Palgrave Macmillan.

Turner J and Moran D (2015) Carceral Geographies VII: Future Directions in Carceral Geographies panel session at the Annual Meeting of the American Association of Geographers, Chicago, IL.

Part I

Tension

v. a constrained condition that results when forces act in opposite directions away from each other

2 Mobile carceral logics

Aboriginal communities and asylum seekers facing enclosure in Australia's Northern Territory

Kate Coddington

The news that an Aboriginal man died while in prison in Alice Springs, Australia, in June 2015 represented just the latest tragedy in a long series of Aboriginal deaths in custody in Australia's Northern Territory; indeed, another Aboriginal man from Yuendemu had died while incarcerated in Darwin just a few weeks earlier (Davidson, 2015). In the same communities where Aboriginal children were once separated from their families and confined within institutions like Darwin's Kahlin Compound, or Retta Dixon Home, disproportionate rates of incarceration work to separate Aboriginal families once again. Yet the Northern Territory has also become known for a different sort of contemporary incarceration: the mandatory detention of asylum seekers in as many as six different detention facilities across the Territory. By 2011, the Territory's capacity to detain over 3,200 asylum seekers made it known as the 'Detention Capital of Australia'.

In this chapter, I argue that the Northern Territory's disproportionate trends in Aboriginal incarceration and its simultaneous embrace of migrant detention represent the mobility of particular 'carceral logics' at work. By carceral logics, I mean the underlying rationales that suggest carceral 'fixes' to contain what are perceived as social problems: for example, the logic that constructs mandatory detention as a 'solution' to the perceived 'problem' of asylum seekers with unknown backgrounds and identities. I argue, however, that carceral logics move through time and space, reconstituting historical reliance on confinement as a response to Aboriginal community endurance in different forms to adapt to the contemporary context.

To explore the *mobility* of carceral logics in Australia's Northern Territory, I focus on two parallel sets of policies that promote carceral 'fixes' to contemporary social issues. The first is the policy of mandatory detention of asylum seekers starting in the 1990s (in both on- and offshore facilities), and the second is the 2007 Northern Territory Emergency Response (NTER) legislation that exacerbated trends of confinement for Aboriginal people. The two sets of policies are chosen to highlight different aspects of contemporary Australian reliance on confinement and imprisonment as a response to difference, particularly difference that challenges Australian national identity and senses of belonging, yet both are also grounded in the historical spaces of enclosure that underpinned Australian settler colonial practices in the Northern Territory. I suggest that carceral logics manifest themselves differently in these two sets of policies, but that, as I detail below, both have a common underlying reliance on a constellation of particular elements: the racialised Other, the decontextualised problem, and the carceral as a method of obscuring the existence of the social problem itself. I argue that these elements give rise to carceral spaces such as the prison or the detention centre, but that these underlying logics are also mobile. Their mobility allows for the expansion of these carceral spaces into new areas of people's everyday lives, resulting in increasing levels of confinement, containment, and enclosure.

To explore the impacts of carceral logics in Australia's Northern Territory, I carried out a combination of participant observation, 25 semi-structured interviews, and archival and secondary source research in Darwin between November 2011 and March 2012. I used participant observation to learn about community dynamics in Darwin, including those occurring during personal visits to detention centres and meetings of advocacy groups. I also conducted 25 semi-structured interviews with advocates, local historians, and other interested parties. Archival research at the Northern Territory Parliamentary Library and the Charles Darwin University Library helped to provide historical and geographical context for policy development. Together, the variety of sources allowed me to focus and prioritise research findings, triangulate them for greater internal consistency, and juxtapose the histories of Aboriginal communities and detained asylum seekers to highlight the carceral logics affecting both populations.

To develop this argument, I first summarise the two sets of public policies taken up in this analysis. Next, I theoretically situate my understanding of *how* carceral logics move by drawing on literatures related to policy transfers and the 'carceral turn'. I then explore the mobility of carceral logics within the two public policies directed at Aboriginal residents of the Territory and asylum seekers, tracking their movement through time and space as well as how they extend carceral space beyond the space of the institution. Finally, I conclude by stressing the violence inherent in the movement of carceral logics, a violence that language such as 'policy transfers' often obscures.

Australian policies of enclosure

Australia's use of mandatory detention for asylum seekers arriving by boat dates back to 1991, when increasing unease about boats of Cambodian asylum seekers prompted the government to open Port Hedland Reception and Processing Centre. The Migration Amendment Act (1992) authorised mandatory detention for unauthorised boat arrivals. The immigration department contracts the daily operation of detention facilities to a changing array of multinational subcontractors such as SERCO and Transfield Services. The Australian public began to take more notice of the conditions of detention in 1999 after the construction of Woomera detention centre, but the case of the Norwegian freighter *Tampa* brought asylum seekers fully into the limelight.

In late 2001, a ship carrying over 400 asylum seekers departed Indonesian shores. After the ship began to sink, the *M.V. Tampa* rescued the 433 passengers. The captain attempted to deliver them to Australian-owned Christmas Island, but the Australian government refused to let them ashore for several days. The Howard government 'drew a line in the sea', as Perera writes (2002: 1), and refused entry to the asylum seekers, ignoring the distress calls of the Norwegian captain. That same evening, Prime Minister Howard rushed the Border Protection Bill through parliament, legislation that radically changed the Australian landscape for asylum seekers. The bill retroactively excised offshore territories for the purposes of migration claims, and established a dual system of asylum processing, with a truncated refugee claims process and limited access to legal services or judicial review (Perera, 2002). The legislation authorised both the interception of asylum seekers arriving by boat by the Australian military and their diversion to Pacific island nations for processing as part of the 'Pacific Solution'.

Since the 1990s, detention has become a major part of Australian border enforcement. Offshore processing of asylum seekers occurs at detention centres on Nauru and Papua New Guinea. Yet onshore detention also continues: as of August 31, 2015, 2,763 migrants were detained onshore, representing over 60 per cent of all migrants detained at Australian facilities throughout the region (ASRC, 2015). Onshore detention facilities located in remote and inaccessible areas demoralise asylum seekers and create or exacerbate mental-health issues. Incidents of self-harm, frequent riots, hunger strikes, arson, suicide, jumps from roofs, or protests by sewing lips together characterise asylum seekers' experiences in detention. Australia continues to try to limit public awareness of conditions of detention, and in 2015 passed the Border Force Act, which threatens potential whistleblowers with a 2-year jail sentence if they disclose information about the operation of detention facilities (Doherty and Farrell, 2015).

The 2007 NTER legislation was proposed in a similar context of a perceived crisis, this one occurring in Aboriginal communities, amidst escalating rhetoric that echoed the framing of the arrival of the *M.V. Tampa* in 2001. Debates over the 'viability' of Aboriginal communities in the Northern Territory had been gaining intensity since the 1970s, as federal policies of assimilation were replaced with self-determination policies that encouraged, but only minimally funded, Aboriginal people's resettlement on traditional lands (Kowal, 2008). Yet self-determination policies increasingly became the subject of public debates, and media and public figures began publicly attributing community distress to Aboriginal culture itself. Sensational media accounts of welfare dependency, corruption, substance abuse, and violence culminated in May 2006 with ABC's *Lateline* reporting sexual abuse in Aboriginal communities using graphic descriptions never before heard during primetime Australian television (Pether, 2010: 26; Watson, 2011: 911). The media explosion eventually prompted the creation of a board of inquiry in 2006 to investigate claims of Aboriginal child sexual abuse.

On June 15, 2007, the board of inquiry released the Ampe Akelyernemane Meke Mekarle, or 'Little Children Are Sacred' report, carefully noting that 'abuse of children is not restricted to those of Aboriginal descent, or committed only by those of Aboriginal descent, nor to just the Northern Territory' (Wild and Anderson, 2007: 5). While the report gestured towards the colonial context of Aboriginal community disintegration, the resulting media and government debate over the report instead characterised Aboriginal people as perpetrators of 'problem sexual behaviour' (Pether, 2010: 31). Six days later, Prime Minister Howard announced a sweeping legislative package to address the 'national emergency' regarding the situation of Aboriginal children in the Northern Territory. A minister described the government's plan as a response to Aboriginal communities that had become 'failed societ[ies] where basic standards of law and order and behaviour have broken down' (Watson, 2011: 912).

The NTER legislation included 'law and order' measures, financial controls over Aboriginal Australians, and control over Aboriginal lands, measures that extended far beyond the report's recommendations. Aboriginal communities were designated as 'Prescribed Areas' where the possession of alcohol or pornography would be forbidden. The federal police were granted new authority, and courts were also forbidden to consider customary laws when setting bail or issuing jail sentences. The NTER disbanded community employment programs and delegated community service provision to new business managers. The NTER established welfare quarantine for Aboriginal Australians' public benefits, allowing only the purchase of 'essentials' by means of a debit card. Community lands were at risk as well: the legislation loosened regulations governing access to Aboriginal land and the government acquired compulsory 5-year leases of traditional lands (Watson, 2011). To apply the measures of the NTER *directly* to the Aboriginal communities of the Northern Territory, the legislation also lifted the protections of the Racial Discrimination Act 1975. Despite widespread community protests in the Northern Territory and throughout Australia, the provisions of the NTER were renewed in 2012, and the welfare quarantine programme was extended to the whole of the country in 2009, targeting vulnerable populations (both Aboriginal and non-Aboriginal) receiving government support.

The mobility of carceral logics

Carceral logics underpinning policies directed at Aboriginal communities and asylum seekers become mobile in different ways. First, I explore the mobility of carceral logics as a type of policy transfer: the movement of the carceral 'solution' from past sets of social problems to contemporary ones, and their movement from one set of social problems to another. I argue that carceral logics underpinning the history of institutionalisation and imprisonment of Aboriginal Australians in Australia's Northern Territory became reconfigured under the NTER, reconstituting carceral 'solutions' within contemporary Aboriginal community life. At the same time, however, I also argue that carceral logics underpinning Aboriginal policies also morphed to address a different social 'problem:' that of the unknowable asylum seeker. Together, these carceral 'fixes' result in incarceration of various forms: the disproportionate imprisonment of Aboriginal populations as well as the detention of asylum seekers. Yet I argue that the mobility of carceral logics is not just a matter of policy mobility through history or across the network of contemporary Australian social issues, but also a matter of the mobility and expansion of carceral spaces, extending the carceral 'fix' of imprisonment and detention into less tangible forms of enclosure and confinement.

Ong's (2007) understanding of the movement of neoliberal logics illuminates just how such mobile logics might operate. For Ong (2007), neoliberal logics of governing migrate and are mobilised differently in different contexts. Rather than a process of replication, this movement becomes a process of reconfiguration, an assemblage shaped by the particular dynamics of governance and the interplay between them. Peck and Theodore (2010) outline a similar process, where the mobility of specific policies takes shape as a constant evolution or transformation. Particular policy pieces travel and become differently aligned depending on the context in which they arrive. Like these policies, carceral logics reproduce in a non-linear fashion, mutating rather than replicating, yet drawing on some of the same established discourses. For example, the racialised social 'problem' of the Aboriginal or the asylum seeker generates insecurity about Australian national identity or belonging. Next, processes of individualisation strip that racialised subject from its social and political context, such as the contemporary war on terror that threatens the lives of asylum seekers from Afghanistan and Iraq, or the complex history of colonization robbing generations of Aboriginal families of security and livelihoods. Finally, the carceral 'fix' then obscures these problems both ideologically, by criminalising problematic populations, and physically, by incarcerating them far from public view.

De Lissovoy (2012: 739) places race centrally within the global expansion of carceral logics, what he calls the 'carceral turn' based on the 'excessive' expansion of the prison industrial complex. He argues that racialised repression represents an essential political economic strategy within neoliberal capitalism. Older racist logics underpin new neoliberal strategies of repression, and these older racist logics explain why it is that states turn to prisons to 'fix' structural problems of the state – these new strategies represent an intensification of pre-existing logics, rather than a novel neoliberal strategy (De Lissovoy, 2012). As carceral logics move through time and space, therefore, they continue to rely on the same essential discourses that have justified their existence in the past. In Australia's settler colonial context, scholars have drawn explicit connections between past practices of Aboriginal and migrant enclosure (Perera, 2002), linking especially the unreconciled nature of colonial dispossession with contemporary struggles over asylum seeker arrivals (Tedmanson, 2008). Both populations are explicitly racialised. In the case of Aboriginal Australians, the foundational legal principle of *terra nullius*, or the empty continent, became the justification for the colonization of Australia, based on the early settlers' belief that Aboriginal people would perish quickly when faced with modern life. In the case of asylum seekers, they too become racialised through the dual system of asylum processing, which permits most asylum seekers arriving by plane to bypass the system of mandatory detention required for 'irregular maritime arrivals'. These categories are highly racialised and classed: those who can afford plane fare, visas, or convincing false documentation enter by airplane, bypassing the detention system. The racism that underpins mandatory detention only for boat arrivals, Giannacopoulos (2011: 4) argues, is the 'product of this colonial system and not simply the product of bad laws'.

The racialised problematisation of Aboriginal or asylum seeker populations combines with the individualised rationale of blame that strips people from their social context and the history of racialised, carceral strategies that make up Australian policy toolkits. These elements thus form a reconfigured policy assemblage that constructs the carceral as a method of addressing new, contemporary social ills. Disproportionate imprisonment of Aboriginal populations and detention of asylum seekers result, as I detail in the next section. Yet carceral logics produce expansive carceral spaces that go beyond the space of the institution. As Gill et al. (2013) write, carceral spaces are not just prison spaces, but represent forms of confinement that extend carceral effects into new terrain. Moran et al. (2013: 121), for example, describe how prisoner transport smooths over the distinctions separating the outside and inside of prison, arguing for a vision of carceral space as porous, and developing a 'more complex and nuanced notion of the "carceral". as mobile, embodied, and transformative'. As I detail below, part of the impact of mobile carceral logics is also to extend the space of the carceral beyond the walls of the institution, expanding processes of racialised confinement into other parts of Aboriginal Australians' and asylum seekers' daily lives.

Carceral logics at work

Incarceration, particularly the imprisonment of white Australian convict settlers, plays a central role in the contemporary Australian national imaginary. Yet contemporary turns toward carceral solutions in response to perceived social problems draw much more directly on Australia's long history of incarcerating Aboriginal Australians. Settlers initially had no formal policy towards Aboriginal people, believing that they would quickly die off, but Aboriginal community resistance efforts led to tactics such as hunting parties to kill Aboriginals for sport (Tedmanson, 2008). By the mid-nineteenth century, Australia created protectorates that took 'custodianship' of all Aboriginal children, and enacted legal segregation on reserves, which became sites of constant surveillance, control over mobility, and forced labour. Offshore islands became forced labour prison camps or quarantine stations (Tedmanson, 2008). Contemporary reconfigured practices of enclosure build on these racialised practices of imprisonment (Bashford, 1998). Together, policies directed at Aboriginal communities relied on multiple forms of the carceral to address the social ills that Aboriginal communities represented, from unwanted mobility or poor parenting to the spread of disease.

While the authors of the 'Little Children Are Sacred' report had contextualised child sexual assault within the larger context of community unravelling brought about by the history outlined above, the government interpretation of Aboriginal communities as 'failed societies' clearly stripped communities from this colonial context. The combination of the racialised social 'problem' of the Aboriginal population in the Northern Territory, the process of removing this population from its social and political context, and the turn towards the carceral 'fix' as a solution for Aboriginal social problems demonstrates the mobility of carceral logics across time. The carceral 'fix' exemplified by historical practices of colonial enclosure is carried forward in time. The carceral 'solution' to problems within Aboriginal communities becomes reconfigured as the disproportionate imprisonment of Aboriginal populations, and the exacerbation of this process under the NTER.

Aboriginal people are disproportionately represented in the Australian criminal justice system. Overall, Australia imprisons a small share of the population (166 per 100,000 adults in 2010) but incarcerates Aboriginal people at the much higher rate of 2,247 per 100,000 (Guerino et al., 2010: 1; Australian Bureau of Statistics, 2012). Men represent the majority of Indigenous[1] people in prison. For a point of comparison, at the end of the apartheid regime in 1993, South Africa had an imprisonment rate for black men of 851 per 100,000; Australia has a rate of 4,194 per 100,000 (Ting, 2011: 2; Australian Bureau of Statistics, 2012). Aboriginal and Torres Strait Islanders lead in nearly every indication of disproportionate imprisonment, and are more than 20 times as likely as non-Indigenous people to be in jail without a sentence, 26 times more likely to be jailed after sentencing, and more likely to be in prison after a prior imprisonment (Australian Bureau of Statistics, 2011).

Moreover, implementation of the NTER only strengthened the already disproportionate levels of incarceration in the Northern Territory. General levels of imprisonment are already higher in the NT than in other parts of Australia. Indigenous prisoners represent 82 per cent of the NT prison population (Australian Bureau of Statistics, 2011: 27). Compared with other states, the NT has the highest overall imprisonment rate, at 762 per 100,000, and the overall prison numbers are rising faster in the NT than anywhere else in Australia (Australian Bureau of Statistics, 2011: 20). The rapid increase in prisoners has prompted the construction of a new 1,000-bed jail near Darwin that was predicted to be over capacity as soon as it opens (Aboriginal and Torres Strait Islander Social Justice Commissioner 2009).

High rates of imprisonment are related to differential police treatment of Aboriginal communities, as well as obstacles Aboriginals face both within and beyond the criminal justice system. Policing in the Northern Territory, for example, targets homes and communities on Aboriginal reserves and simultaneously employs clean-up or tidiness initiatives to discourage Aboriginal gatherings in public spaces (Cunneen, 2001). Under the NTER, police presence increased with new police stations and intensified policing practices, including unrestricted searches of Aboriginal homes and cars (Pilkington, 2009). Aboriginal people are more likely to be jailed – rather than summoned – for minor offenses, like public order or suspended drivers' licenses, which incur bail conditions and charges associated with failure to meet bail. Furthermore, if rearrested, Aboriginal people tend to face harsher penalties because of prior jail sentences (Cunneen, 2001; Ting, 2011). Mandatory sentencing laws, implemented throughout the 1990s, limited judicial discretion for sentences (Ting, 2011). Prosecutors are often blamed as being 'soft' on Aboriginal people, and despite the numbers of studies showing the disproportionate rates of Aboriginal sentencing, media and commentators portray Aboriginal people as thinking of jail as a vacation, inherently or innately violent, or as incapable of understanding consequences of violent actions (Cunneen, 2001).

Yet other challenges extend beyond the nature of the criminal justice system. The correctional system also tends to strip people from their cultural and socio-economic contexts, including collective trauma, colonization, forced removal of children, and widespread institutionalisation, all of which have a significant impact on incarceration rates. Health and mental health also play a significant role in the criminalisation of Aboriginal people, as do differences in body language and English comprehension (Cunneen, 2001). Aboriginal people also engage in resistance to policing and discriminatory laws, and what the criminal courts interpret as juvenile 'delinquency' or 'justice offenses' may often represent historical forms of Aboriginal resistance such as passivity, non-cooperation, or absconding (Cunneen, 2001).

The NTER promises to continue to increase levels of Aboriginal imprisonment even further. Both the 2007 legislation and the follow-up 2012 Stronger Futures bill have toughened bail conditions, decreased the rate of acceptance of applications for parole, increased levels of policing leading to jail sentences, and maintained the NT's record of extremely low rates of community-based (non-custodial) sentences and alcohol and sexual abuser rehabilitation programs (Aboriginal and Torres Strait Islander Social Justice Commissioner, 2009: 43). The NTER also prohibits taking into account customary law for sentencing, effectively barring the use of Aboriginal law in NT courts. Not only does this measure increase the length of sentences, it also represents further neglect of Aboriginal people by the Australian legal system. Disproportionate imprisonment is one way in which carceral logics have moved from the past to the present, continuing to confine Aboriginal populations at higher and higher levels in an effort to gain control over the 'failed societies' they represent.

Carceral logics beyond imprisonment

Yet the enclosure encouraged by the passage of the NTER is not limited to increased levels of Aboriginal incarceration. The extension of carceral spaces beyond the prison represents another way in which carceral logics are mobilised, as carceral 'fixes' expand into areas of Aboriginal people's daily lives. The NTER provides a clear case of a racially discriminatory policy explicitly targeting, and criminalising, Aboriginal spaces beyond the prison. Indeed, the Howard administration had to lift the Racial Discrimination Act 1975 to implement the legislation. Aboriginal people are subject to the NTER simply because they are Aboriginal. For example, in an encounter reported at a Centrelink (welfare benefits) office in

Galiwin'ku, Elcho Island, John assisted his mother Julie in determining why she was subject to welfare quarantining, or income management.

> The officer answered, 'It's a response to the Little Children are Sacred Report'. John was surprised, 'You must think she is a child abuser. I want my mum exempted from income management.' The officer asked, 'What are the reasons she should not be income managed?' John thought, then demanded, 'First you tell me the reasons she is on it.' At first the officer could not answer; eventually he replied, 'Because she lives on Aboriginal land'. (Webb, 2008: 18)

New regulations over Aboriginal land became one way in which carceral logics were extended beyond the space of the prison. The NTER created a new designation for more than two-thirds of the Aboriginal communities in the Northern Territory. As 'Prescribed Areas', these communities were subject to legal enclosure, governed by new restrictions on alcohol, pornography, and computer use. Prescribed Areas were subject to informal social enclosure as well through the use of large signs calling attention to alcohol and pornography restrictions. Aboriginal communities targeted with Prescribed Area regulations were isolated and defined as 'problem' spaces, representing the sites of pathologised cultural failings, exemplifying how the 'fix' of the carceral space moved from formal spaces of imprisonment to much more informal, yet effective forms of enclosure.

Other methods of informal enclosure have resulted from the NTER legislation. The complications of administering income management programs, for example, tie Aboriginal people to Centrelink (welfare benefits) offices even as they need to undertake travel for funerals or other significant occasions (Coddington, field notes, February 24, 2012). Women in Darwin's Bagot community reported other forms of enclosure they experienced while having income quarantined. Initially, the restrictions on funds prevented them from paying for taxis or public transit, cutting off their access to groceries and basic household supplies, even in Darwin's urban centre (Coddington, field notes, February 24, 2012).

Another example of the creation and policing of problematised spaces is the experiences of Aboriginal people 'sleeping rough', known as long grassers, a population that has increased under NTER policies. Extensive policing and continued depictions of Aboriginal inhabitants as 'out-of-place' have resulted in informal enclosure and spatial segregation of the long-grasser population. When questioned about the increased levels of Aboriginal people sleeping rough in Darwin after the NTER, hundreds of non-Aboriginal Darwin residents surveyed attributed sleeping in the long grass to romanticised understandings of Aboriginal 'walkabouts'. No one, the study reports, 'identified dispossession, internal displacement, colonization, stolen generation, cultural genocide or any other government policy geared toward the control of Aboriginal people as a reason that they might be staying in the long grass' (Holmes and McRae-Williams, 2008: vi). Carceral logics move across time, becoming reconfigured under the NTER, resulting in the expansion of carceral spaces beyond prison walls.

Carceral logics directed at asylum seekers: Detention and beyond

Australian policies develop over time a reliance on the carceral as a means of 'fixing' perceived social problems involving racialised populations. Policies evolve, transform, and move over time, yet they share an underlying *logic* that reads problems through the framework of colonial practices of enclosure: the racialised others, the abstraction of their lives out of social and political context, and the reliance on the carceral as an ideological and practical means for concealing the issues away from public view. Even as carceral logics were being mobilised in new ways in Aboriginal communities, so too were they being used to engage with problematic populations of asylum seekers arriving to Australia by boat. Asylum seekers today face similar policies of enclosure that range from incarceration, in the form of mandatory detention, to the individualised forms of self-surveillance that occur long after the asylum seeker is released. Like policies directed at Aboriginal communities, however, carceral policies directed at asylum seekers *also* build on long history of turning to the carceral to deal with unwanted immigration to Australia, including the 1901 Immigration Restriction Act and public health quarantines directed at migrants (Bashford, 1998).

Migrant detention is a paradoxical institution: on the one hand, practices of detention rationalise and normalise processes of criminalisation in public discourse. Members of the public may come to believe that migrants must *be* criminals if they are detained like criminals. In reality, however, the securitisation of migration through enclosure practices like detention is itself a way of perpetuating *in*security. More people become threatening all the time, yet the source of their threat is uncertain: they are 'presumed to be dangerous in a non specific way' (Bashford and Strange, 2002: 520). Detention practices designed to shield the Australian public from the potential danger of the asylum seeker are deeply traumatising for asylum seekers themselves, especially as migrants stay in detention facilities for lengthy and indefinite periods of time. Isolation and the remote locations of detention facilities demoralise asylum seekers and create or exacerbate mental-health issues, which often continue long after they are released. Detention also isolates asylum seekers from communities of support (Perera, 2002).

Yet the effects of enclosure affect asylum seekers even after their release from detention, expanding the space of the carceral beyond the walls of the detention centre. Migrants internalise the need to surveil and discipline themselves, carrying a sense of insecurity with them wherever they go (McDowell and Wonders, 2009). Contemporary policy trends in Australia reinforce this sense of insecurity. Starting in 2013, asylum seekers living in the community were required to sign a 'Code of Conduct' stating they would cooperate with all immigration department requests and would not engage in antisocial or illegal sexual behaviours (Taylor and Laughland, 2013). Asylum seekers who are found in violation of the code risk having their very low income support benefits curtailed or their visas cancelled, leading to detention and deportation. The ongoing insecurity of self-surveillance that asylum seekers practice as they live under the Code of Conduct suggests how for asylum seekers, too, the space of the carceral 'fix' is not limited to the prison, but becomes entangled with their mobile bodies over time as well.

The reintroduction of Temporary Protection Visas (TPVs) in 2015 represents another type of ongoing insecurity for asylum seekers. The TPV allows asylum seekers to live in Australia for three years, but asylum seekers can be returned to their country of origin whenever Australian officials deem it safe, placing them in an extended period of limbo without possibility of permanent protection. The current TPV application process has also been 'fast-tracked' to determine asylum seeker fates with very limited means for appealing decisions, what Refugee Council of Australia Chief Executive Paul Power called a 'shattering blow for asylum seekers who face the grave risk of being returned to danger' (RCOA, 2014). Temporary status brings permanent insecurity for migrants, who continue to face the effects of policies of enclosure even after they are released from detention facilities. Mobile carceral logics construct both detention centres and the carceral spaces beyond, such as the dangerous limbo of the temporary protection visa, as 'fixes' to the problem of unwanted, racialised migrants, mobilising both historical and contemporary uses of the carceral in new destructive ways.

Conclusions

In this chapter, I have argued carceral logics move across space and time. Australia's history of settler colonialism established the 'fix' of the carceral institution as a response to the social 'problem' of Aboriginal survival. Carceral logics brought together three key elements: a problematised, racialised population; the abstraction of that population from their social and political context; and the carceral space as a 'fix' to shield the problem from public view. Carceral logics moved through time, becoming reconstituted under the NTER to position the carceral as a response to the perceived social problems of contemporary Aboriginal community life. Carceral logics also moved across policy arenas to address the 'problem' of the unknowable asylum seeker. In each case, the carceral spaces built upon carceral logics involved more than simply spaces of imprisonment and detention, but expanded the reach of the carceral far into people's everyday lives.

The movement of carceral logics in Australia demonstrates the importance of policy mobility not simply across space, but also through time. As a settler colonial nation founded through practices of violent enclosure, Australian understandings of belonging and national identity have been woven through with carceral logics since settlers arrived. The historical potency of carceral logics in Australia, especially those designed to address the 'problems' of racialised populations, helps to explain the pull of these logics in the present. The historical underpinnings of contemporary policies of enclosure such as disproportionate imprisonment, Prescribed Areas, and the detention of asylum seekers also demonstrate the importance for thinking about mobility as both a spatial *and* temporal reconfiguration.

Peck and Theodore (2010: 173) rightly criticise contemporary debates about policy transfer as being focused on rational actors and the replication of best practices across new sites and circumstances, highlighting instead critical approaches that 'explore … processes of networking and mutation'. Yet this analysis of policy mobility remains curiously disembodied, detached from the real-world consequences of the mobility of particular logics and regimes of governance. A focus on the everyday consequences of policy mobility highlights instead the structural *and* everyday forms of violence that characterise logics of contemporary governance such as the carceral logics governing policy-making in Australia's Northern Territory. From everyday forms of confinement such as self-surveillance and income quarantine to the violence of prison and detention, asylum seekers' and Aboriginal Australians' experiences suggest that attention towards the mobilities of policies needs also to be directed at their often violent social consequences, violence often obscured by the clean language of 'networks,' 'mutations', and 'policy transfers'.

Note

1   Indigenous refers to both Aboriginal and Torres Strait Islander communities who are often grouped within Australian statistics.

References

Aboriginal and Torres Strait Islander Social Justice Commissioner (2009) *2009 Social Justice Report*. Canberra: Australian Human Rights Commission.

ASRC (Asylum Seekers Resource Centre) (2015) *Statistics*. Available at: www.asrc.org.au/resources/statistics/.

Australian Bureau of Statistics (2011) *Prisoners in Australia, 2011*. Canberra: Australian Bureau of Statistics.

Australian Bureau of Statistics (2012) *Corrective Services March Quarter 2012*. Canberra: Australia Bureau of Statistics.

Bashford A (1998) Quarantine and the imagining of the Australian Nation. *Health* 2(4): 387–402.

Bashford A and Strange C (2002) Asylum-seekers and national histories of detention. *Australian Journal of Politics and History* 48(4): 509–527.

Cunneen C (2001) *Conflict, Politics and Crime: Aboriginal Communities and the Police*. Crows Nest, NSW: Allen & Unwin.

Davidson H (2015) Aboriginal man's death in custody in Northern Territory prompts scrutiny. *The Guardian*, 21 June.

De Lissovoy N (2012) Conceptualising the carceral turn: Neoliberalism, racism, and violation. *Critical Sociology* 39(5): 739–755.

Doherty B and Farrell P (2015) New inquiry into detention centres will allow whistleblowers to give evidence. *The Guardian*, 12 October.

Every D and Augoustinos M (2007) Constructions of racism in the Australian parliamentary debates on asylum seekers. *Discourse and Society* 18(4): 411–436.

Giannacopoulos M (2011) Nomophilia and Bia: The love of law and the question of violence. *Borderlands e-journal* 10(1): 1–19.

Gill N, Conlon D and Moran D (2013) Dialogues across carceral space: Migration, mobility, space and agency. In: Moran D, Gill N and Conlon D (eds) *Carceral Spaces: Mobility and Agency in Imprisonment and Migrant Detention*. Ashgate: Farnham, 239–248.

Gosford B (2012) Broken promises and 'dumb insolence': The death of Kwementyaye Briscoe. *Crikey*. The Northern Myth: 1–4.

Guerino P, Harrison PM and Sabol WJ (2010) *Prisoners in 2010*. Washington, DC: US Department of Justice.

Holmes C and McRae-Williams E (2008) *An Investigation into the Influx of Indigenous 'Visitors' to Darwin's Long Grass from Remote NT Communities*. Hobart: National Drug Law Enforcement Research Fund.

Kowal E (2008) The politics of the gap: Indigenous Australians, liberal multiculturalism, and the end of the self-determination era. *American Anthropologist* 110(3): 338–348.

McDowell MG and Wonders NA (2009) Keeping migrants in their place: Technologies of control and racialised public space in Arizona. *Social Justice* 36(2): 54–71.

Moran D, Piacentini L and Pallot J (2013) Liminal transcarceral space: Prisoner transportation for women in the Russian Federation. In: Moran D, Gill N and Conlon D (eds) *Carceral Spaces: Mobility and Agency in Imprisonment and Migrant Detention.* Ashgate: Farnham, 109–125.

Ong A (2007) Neoliberalism as a mobile technology. *Transactions of the Institute of British Geographers* 32(1): 3–8.

Peck J and Theodore N (2010) Mobilising policy: Models, methods, and mutations. *Geoforum* 41(2): 169–174.

Perera S (2002) A line in the sea. *Australian Humanities Review* September: 1–8.

Pether P (2010) Reading the Northern Territory 'Intervention' from the margins: Notes toward a feminist social psychoanalytic ethics of governmentality. *Australian Feminist Law Journal* 33(1): 19–36.

Pilkington J (2009) *Aboriginal Communities and the Police's Taskforce Themis: Case Studies in Remote Aboriginal Community Policing in the Northern Territory.* Darwin, NT: NAJAA & CAALAS.

Prout S and Howitt R (2009) Frontier imaginings and subversive indigenous spatialities. *Journal of Rural Studies* 25(4): 396–403.

RCOA (Refugee Council of Australia) (2014) *Asylum Laws Will Fast-Track Vulnerable People to Danger*, 5 December. Available at: www.refugeecouncil.org.au/media/asylum-laws-will-fast-track-vulnerable-people-to-danger/.

Taylor L and Laughland O (2013) Asylum seekers living in Australia forced to sign code of conduct. *The Guardian*, 16 December.

Tedmanson D (2008) Isle of exception: Sovereign power and Palm Island. *Critical Perspectives on International Business* 4(2/3): 142–165.

Ting I (2011) Aboriginal crime and punishment: Incarceration rates rise under neoliberalism. *Crikey*, 15 December.

Watson N (2011) The Northern Territory Emergency Response: The more things change, the more they stay the same. *Alberta Law Review* 48(4): 905–918.

Webb R (2008) The Intervention – A message from the Northern Territory. *Indigenous Law Bulletin* 7(9): 18–21.

Wild RSL and Anderson P (2007) *Ampe Akelyernemane Meke Mekarle* [Little Children Are Sacred] *Report of the Northern Territory Board of Inquiry into the Protection of Aboriginal Children from Sexual Abuse.* Darwin: Northern Territory Government.

3   The ambivalent camp

Mobility and excess in a quasi-carceral Italian asylum seekers hospitality centre

Roberta Altin and Claudio Minca

The Gradisca camp

Gradisca, Italy: 3.30 p.m., 28 May 2015. The documents stating the authorisation released by the Gorizia Prefecture in our hands, together with our identity cards, we ring the bell of the Gradisca Hospitality Centre for Asylum Seekers, the CARA (*Centro Accoglienza Richiedenti Asilo*). The centre features high concrete walls together with security cameras inside and outside the centre. We enter 'the camp', after being subjected to the metal detector and the severe check-up of our documents by the guard sitting behind a glass barrier. The feeling is that of entering a highly securitised 'danger zone'. Once the documents are verified – and retained by the guard – we pass a second security door to be faced by two internal routes: the first, leading to the former CIE (*Centro Identificazione ed Espulsione* – that is, Centre for Identification and Expulsion) formally closed two years ago; the second, to the green areas of the CARA, hosting individuals who have applied for the status of refugee. Gates, barriers, bars, and barbed wire 'protect' the entire compound: the only entry/exit point is the double security door we have just passed. There is another gate for vehicles, equally subjected to strict military check-ups. This entry procedure immediately reveals the quasi-carceral regime of mobility characterising this site of 'hospitality'.

The Gradisca centre was inaugurated in 2008 by converting a former military compound which was part of a key defensive line along the border with Yugoslavia (now Slovenia) during the Cold War decades. The reuse of this site was a by-product of the introduction in the Italian legislation by the conservative Berlusconi cabinet of the crime of '*clandestinità*' (illegal migration), also known as 'Legge Bossi-Fini' (2002). This decree was in fact followed by a series of 'urgent measures in terms of public security' (L.125/2008), which required the creation of the CIEs, the Centres for Identification and Expulsion. From 2008 to 2013 this former barrack was therefore used as a detention centre hosting foreign individuals accused of illegal behaviour or status (for example, missing the residence permit or identity documents); at the same time, part of the centre was dedicated to the reception of asylum seekers. The CIE closed in 2013 after internal revolts and the vibrant protest of the Gradisca residents opposed to the presence of such a centre on their territory. Currently, this carceral space remains officially closed; however,

Figure 3.1 The entrance of the CARA.

Source: Author's collection.

due to the increasing flow of new migrants along the Balkan route and the related uncertainty concerning the EU policy on migrations and refugees hospitality, such closure is only formal; de facto, in the past year or so, this space has been unofficially reopened to offer provisional accommodation to the exceeding number of asylum seekers reaching the region. We are thus faced with the ambivalent condition of a camp officially aimed at providing hospitality to asylum seekers, which incorporates (and irregularly re-uses) a centre for the expulsion of illegal migrants, that is, a carceral space (MEDU, 2013).

The asylum seekers are formally 'guests' of the centre, which provides them with accommodation and the most essential services while their status is evaluated. However, these guests are subjected to a specific spatio-temporal regime that frames their days, outside and especially inside the centre. This is why we suggest thinking of this centre as a 'camp', that is, a spatial political technology aimed at governing, disciplining, and qualifying the 'migrants', as they are normally described by the Italian media. This hospitality centre/camp, we claim, is a site aimed at segregating and containing individuals with unclear legal status, an attempt to confront the 'refugee problem' by spatialising it into an institution functioning according to a 'quasi-carceral mobility'. By 'quasi-carceral' regime, we intend here a regime based on forms of control and limitation of mobility for subjects who are technically not detainees, since they have committed no crime, but who, because of their status, are nonetheless kept in conditions of partial and selective captivity. Quasi-carceral regimes of mobility allow a certain degree of regulated mobility for those subjected to them; however, their spatialities are all too often structured around former carceral institutions, and adopt forms of disciplines of a carceral kind (see Felder et al., 2014).

In this chapter, we provide critical insights on the mobilities encapsulated in doing research in such quasi-carceral settings. We believe that in the growing literature on mobilities in/of carceral space, there is not enough discussion to date that explicitly reflects on the limited, disciplined, coerced, and power-filled mobilities of both researchers and of the so-called informants. We therefore hope to be able to highlight how this aspect has affected our research. Second, we would like to relate the question of quasi-carceral mobility to the notion of human excess. We thus propose focusing attention to the mobilities of those 'confined' who may temporarily leave the 'camp' and occupy some emerging 'buffer zones' as a form of excess of human life in the city. We accordingly ask whether the quasi-carceral spatialities here discussed are not only about the human excess represented by, literally, people-with-no-status, but also about excess of human mobility/immobility. What we try to argue is that, even though mobilities of migrants are limited – to specific times and spaces – they are also in excess to the normal 'traffic' of people, and this paradoxical 'excess of mobility' is what creates fear among the residents, as it is perceived to threaten the wider social order. In addition, we maintain, during the time spent occupying the camp, migrants and asylum seekers are enfolded within a different notion of excess – an excess of time that holds them immobile, or unable to move beyond their current situation.

We do so by indirectly engaging with recent work on 'carceral geographies' (see Moran, 2012, 2015; Moran et al., 2013; Turner, 2012, 2013a, 2013b; Peters, 2014) and work on refugees mobility calling for a closer examination of the related spatialities and micro-politics (see, among others, De Genova and Peutz, 2010; Gill, 2009; Malkki, 1995; Mountz, 2011; Verdirame and Harrel Bond, 2005). Since this literature is largely examined in the introduction to this collection, we do not find it necessary to recall it again here. This project, developed in 2014–2015, was largely based on participant observation inside and around the camp, textual analysis of official documents and the related media coverage concerning the CARA, and a series of interviews. The interviews have involved asylum seekers hosted at the CARA, CARA staff members, and some key local authorities – in particular the former and the present Mayor of Gradisca. The ambivalence between the undefined status of the asylum seekers and the spatio-temporal disciplinary framework 'hosting' them is at the origin of specific camp geographies that include some of the surrounding areas occupied by these subjects-in-waiting during their 'free time', where they suddenly become visible to the local residents of Gradisca as 'people in excess'. The chapter discusses these ambivalent geographies to show how they are fundamental in the production and the management of a docile and undefined subject, the asylum-seeker-in-waiting, a figure central to the politics of the contested hospitality centre located in this border region.

Inside the camp

We are initially approached by the vice-Prefect, who provides us with a description of the camp and its functions, to be then assigned to the director of 'Connecting People', the cooperative managing the centre on behalf of the government. The meeting proves difficult and tense: 'we do not like interviews and journalists', the director states at the start, since she believes they have not played a constructive role in commenting the 2013 revolt. During that event, which contributed to the CIE closure, one of the 'rebels' died after falling from the roof where some of the inmates gathered to protest. Notably, in the months following our visit, a series of investigations took to court several managers and staff working for 'Connecting People' accused of misconduct and fraud (as reported in *Il Sole 24 Ore*, [2014]; *Il Piccolo*, [2016], *Il Messaggero Veneto*, [2016]). Apparently, the cooperative had profited from claiming larger than due sums from the government by submitting falsified documents about the costs of the guests' care. At present, another cooperative is in charge of the centre.

The former CIE and the CARA are now separated by numerous barriers and a Plexiglas wall. We are not allowed to visit the former CIE, the official reason being that it is now subjected to renovation work. Meanwhile, we cannot freely roam around the CARA either: as soon as we enter the compound we are accompanied – and directed – by the staff. During the entire visit we are never left alone, neither are we allowed to directly interact with the guests. The staff is kind and, in principle, open to collaborate with academics; however, the visit remains heavily marked by a feeling of diffidence towards us and the related control. Even during the interviews we are constantly 'monitored' by an interpreter (or so-called cultural mediator) working for 'Connecting People'.

The centre hosts today about 260 'migrants', mostly men from Afghanistan and Pakistan, who normally stay for 6–8 months waiting for the procedure of assessment of their status to be completed. The Gradisca 'camp' is located a few kilometres away from the Slovenian border in Gorizia, where the Territorial Commission for the Recognition of the International Protection has its offices. The so-called *prima accoglienza* (first reception) lasts three weeks: at the entrance, the hosts are provided with a kit including clothes and products for their personal hygiene; immediately after that they have an introductory meeting with a 'mediator' to discuss the rules of the camp, its facilities, and the procedure to obtain the status of refugee. In the following days the 'migrants' meet a psychologist and a social worker, while some of the key information concerning each individual is collected, in collaboration with the *Consiglio Italiano dei Rifugiati* (Italian Council for Refugees). The staff then works in close contact with the immigration authorities to obtain the temporary residence permits. Finally, if no particular conditions of psychological vulnerability emerge, the guests have an additional meeting with the 'educators/mediators' during which a tentative CV is drafted and some potential activities are taken into consideration, including theatre laboratories, gardening, the clearing of dismissed areas in the surroundings – tasks typically assigned to prisoners.

What emerges from that specific visit and the interviews is that the CARA guests spend most of their time in a sort of spatio-temporal limbo, since their ordinary life is literally suspended (see also Michalon, this volume). They are allowed to leave the centre only between 8.00 a.m. and 8.00 p.m. After 8.00 p.m. the gates are sealed like in a prison. Exceptional permits are given only by the Prefecture, but requests must be submitted at least three days in advance. The rooms contain from 5 to 8 beds – although in extraordinary moments they may be increased to 10–12. The groups are organised according to nationality; however, this is often not enough to avoid internal tension due to the limited individual space available. The rooms are open to a yard surrounded by barbed wire, but that nonetheless represents a shared space in their internal quasi-carceral condition. The main building includes a cafeteria, three TV rooms, a mosque, and a large aula dedicated to reading and praying. The spaces used by the personnel, the director's office, and the infirmary are located in a separate part of the complex.

Despite our explicit request to interview some of the 'migrants', we are 'advised' not to directly interact with them. The 'mediator' accompanying us thus offers to call a selected sample of 'guests'. The call is made via the internal sound system, with the individuals identified by numbers. Each guest is in fact classified and registered with an identification number based on his arrival time (for example, 151/2015). Guests must always carry with them the card with their respective identification number, since it gives them access to the internal services (like the laundry). The rationale for the selection of the interviewees is not revealed to us and, despite our request to interview five guests, only two attend the meeting. It is not clear whether the other three have changed their programme or have simply refused to be interviewed.

The meeting lasts about one hour, during which we are never allowed to speak directly to the asylum seekers. S.A.J. (25, from Pakistan) and A.M.A. (29, from Afghanistan) provide only vague and politically correct answers. When asked about his motivations to migrate, S.A.J. (28/05/15) declares: 'we have come to Italy to look for safety because at home we risked our lives'. He also claims to be entirely happy with the centre: 'I am satisfied with how they treat us; at the beginning we had to sleep in the street, but only for a short while, since we have been accommodated rather quickly in the CARA' – all in good order, no critique, no issues. Even more emblematic is A.M.A.'s (28/05/15) interview:

> I feel like in a family; I am happy for everything I have received here; I have been in jail for many years back home, I have been vexated in so many ways during my life, and here I have finally found some true hospitality, my status has been recognised, and I am happy.

The interviews seem to suggest the presence of what is at times described as the 'management' of the asylum seekers, according to which the procedure of assessing the refugee status often translates into a period and process in which the guests are desubjectivised by the authorities via numbering, classification, and pervasive forms of standardisation of their memories and self-representations; in other words, instructed to say 'what the institutions like to hear' (Sorgoni, 2011: 53; see also Gill, 2009; Verdirame and Harrel Bond, 2005).

Despite the odd interview settings and our resulting suspicion of the interviewees' self-censorship, when we explicitly ask them what they think of Italian and European politics concerning the 'migrants', they seem rather disoriented and devoid of any sense of the context hosting them:

> We have no contact with Italian people, with the exception of the centre's staff, and the language represents a major barrier ... the Italians see us camped in the fields and think that we may be nomads who sleep on the sidewalk. (S.A.J., 28/5/15)

While walking around the compound, we feel to have entered a tightly controlled space loaded with rules, surveillance, censorship, custody – a spatiality entirely detached from the outside world, deprived of most of the ordinary practices that normally regulate people's daily existence (De Genova and Peutz, 2010). A sort of grand space of exception, where time and social meaning appear as strangely suspended. As noted above, the centre may be visited only after obtaining the authorisation from the Prefecture, which verifies and evaluates the motivations and the status of the applicants. However, even when the authorisation is given, the disciplinary regime in place seriously affects the freedom of movement of the visitors.

A few considerations about the role of the researcher in such quasi-carceral setting are therefore perhaps in order. On the one hand, the authorities are often kind and claim the whole refugee management to be a transparent process, while on the other, the visit tends to be highly constrained: we were never free to roam

Figure 3.2 Outside the camp: 'migrants' walking on the road.

Source: Author's collection.

around, to explore, to speak to the guests. During the visit we are deliberately kept away from the guests who are hanging out, visibly bored, in the garden. We feel deeply regimented, while our own mobility seems entirely controlled by the authorities running the camp, as if they feared that some underlying secret could emerge during our investigation – or as if they considered the 'refugee problem' merely a question of bodies management and the related spatio-temporal disciplining, to be kept away from our scrutiny. Tellingly, photos are not allowed inside and in the immediate surroundings of the camp, fully monitored by the inquisitive eyes of multiple security cameras. Why such a regime of secrecy and containment in a hospitality centre created to help the refugee? From the two limited but still significant rounds of interviews conducted in 2014 and 2015 respectively, what emerged is that the interviewees were not entirely free in interacting with us, their answers clearly being vague and superficial. It became clear in our interaction with 'the camp' that the asylum seekers were subjected to various forms of subtle self-disciplining, in line with the rules and regulations governing the modalities according to which they were assessed and treated, something inducing them to think of themselves and their condition in certain ways: as if, by entering the camp, they had literally incorporated the 'asylum seeker uniform', thus becoming part of the practices and the language that normally accompany the very process of 'qualification' in which they were involved (Ong, 2003).

Furthermore, the interviews and the visit revealed the profound, albeit subtle, forms of co-implication with the 'refugee management' system that such research inevitably implied. From the moment in which we walked into these quasi-carceral realities, we became inevitably complicit in the overall ideological framework at the foundation of the camp. To enter the camp, we have submitted ourselves to 'the rules of the game' and with our very presence implicitly legitimised the apparatus of humanitarian governance there in place. At the same time, the interviewed guests seemed to 'act-as-refugees', that is, to have entirely incorporated the language and the body posture normally attributed (again, by the assessing authorities) to individuals deserving the refugee status. On both sides (the interviewee and the interviewer), what emerged was a pervasive sense of unease, fuelled by the general settings imposed by the camp regime, but also by the reciprocal lack of trust and the ambivalence of the whole process. In such settings, the 'guests' do not trust the interviewers and say as little as possible; in addition, they may use the interview to deliver a specific message to the mediators and the overall institutions (on this see Agier, 2011: 161–174; Sigona, 2014). Instead of being an encounter between two individuals, the interviews tend to take the form of an artificial contact zone between two political categories: the interviewer, who may recognise them as 'victims' or as 'fake asylum seekers'; and the guests, who speak only in line with their desired status. Their life beyond the camp is never mentioned. They embody and enact their condition of subjects-in-waiting. The result is that under such constrained conditions, the whole fieldwork practice seems pervaded by the same spatio-temporal limbo of the camp.

The quasi-carceral mobility characterising this camp is evident in many of its 'devices'. Its horseshoe shape is dotted with separation barriers of all kinds which contribute to producing spatialities aimed at controlling the inmates' mobility and at reducing the possibility of large groups of individuals gathering together, potentially representing a security threat. Initially, 'the CIE was like a cage, with all those oppressive gratings and bars. The then Prefect, Mr. De Lorenzo, obtained from the interior Ministry to replace part of the bars with these plexigas panels' (T.F., former Gradisca Mayor, 17/4/2014). Until its closure in 2013, the CIE could host up to 284 people distributed among three areas marked by different colours depending on the level of risk/danger assigned to each individual: 'We had three areas. One green, one red, one blue. The green was the most peaceful, where they will put individuals considered unproblematic, who were not complaining' (G.F., CARA nurse, 10/3/2014). The best treatment was given to those most collaborative and calm, as it happens in many carceral regimes (Rivoli, 2014). The CIE was characterised by even more specific limitations in the inmates' mobility: for every four rooms, there existed a common area called 'the pool', with a public phone and a vending machine; however, 'no more than four individuals at the same time could leave their room. Accordingly, no more than four people could visit "the pool" at the same time, to avoid problems' (M., CIE/CARA staff member, 21/3/2014).

What was supposed to be an 'administrative detention' had gradually incorporated measures typical of a penal regime of detention, including forms of classification and mobility restriction (Majcher, 2013: 4). As admitted by the Prefect (Z.V., 19/5/2014):

> The objective of the centre was to keep under custody these individuals until their complete identification, however, all too often actions implemented in the name of such procedure responded to different objectives, for example to interventions of social prophylaxis: 'You do not want to be identified? Fine, I will then keep you here. I cannot expel you right away because I do not know who you are and therefore I would not know where to send you to, but in the mean time I keep you locked in'.

The interplay among identification, custody, and exclusion was thus the source of a quasi-carceral spatiality based on a disciplinary regime aimed at producing docile biopolitical subjectivities (Minca, 2005, 2015b; also, Levy, 2010).

During the first set of interviews conducted in 2014 in the CARA, however, some of the guests seemed more open to discuss the custodian regime of the camp. For example, a guest from Pakistan claimed that: '… it feels like a prison, everywhere you turn your head you see gates, bars and barriers. This is definitely not an open space. And the military … I think they are here to control us, but we are completely peaceful' (H., 25, 5/6/2014). Another inmate declared that:

> at times, I feel we are relatively free. However, we are locked in by 8pm, and this is very stressful. The thing is that we have come here to be free and instead we end up in a place where we are imprisoned. This is a true paradox […] At times I go out, but not very often. I spend most of my time here. But this is not home, it is a camp. (J., 23, CARA guest from Nigeria 5/6/2014)

The waiting time within constrained and strictly regulated spaces contributes to create extraordinary temporalities as well (see Conlon, 2011; Armstrong, 2015). For those accommodated at the CARA, time management is part of the disciplinary quasi-carceral spatial condition: the contact with the legal office and the 'cultural mediators', the visits to the psychologist or the infirmary, the Italian language lessons, the refectory, are all overseen by a spatio-temporal regime inspired by carceral institutions. According to a CARA psychologist, 'These rules give a specific rhythm to the everyday life that allows us to have control over these people' (N., 29/5/204). Inactivity, apathy, and lack of stimuli mark a condition described by Dominique Moran (2012: 313) as 'carceral TimeSpace': 'my life here consists of waiting for a better life *outside*' (A., CARA guest, 5/6/2014). None of the interviewees seems capable of planning their future: their existential horizon appears entirely absorbed by the (uncertain) possibility of receiving the status of 'refugee'. This 'identity limbo' therefore translates into a spatio-temporal condition of 'subjects-in-waiting', taken by the obsession of obtaining an interview/audition with the *Commissione Territoriale*, composed of representatives of the Prefecture, the Police, the local authorities and the UNHCR (United Nations High Commissioner for Refugees). For many guests the centre is just a site of temporary accommodation, deprived of any realistic perspective of social integration, where they remain nothing but 'subjects-in-waiting' (Rahola, 2003).

Makeshift camps

The location of the CARA, away from the centre of Gradisca, responds to a broader strategy aimed at isolating and containing 'the migrants' to prevent any interference with the residents' ordinary lives (see Agier, 2008; Pinelli, 2015). According to the former Gradisca Mayor: 'We succeeded in securitising the territory … this area is at once patrolled by the regular police, the military, the border police … the distance from the CARA has helped during the time of the revolt, since it has protected the city centre from vandalism. I trust the position has been helpful, we could not do better than this' (F.T., 17/4/2014). The guests, on the other hand, during the day try to spend as little time as possible inside the centre: 'I try to get out as much as possible. In the morning, I leave, I go out … not to become insane' (S., CARA guest from Afghanistan, 27/5/2014). However, going 'out' often means spending time in an open space along the River Isonzo, a space that the guests normally use for leisure and social interaction and labelled as 'the jungle'. In the jungle, they chat, set up fire, cook, and run a series of social activities. Exactly like in Calais and Melilla, the jungle here is a quasi-abandoned space, a no man's land of sorts (see Altin and Minca, 2017). This is an interstitial spatiality produced by the migrants to complement and resist the quasi-carceral regime of mobility imposed by the CARA, a regime that does not allow conducting an ordinary life, inclusive of elements of privacy and intimacy.

The 'jungle' along the riverbank is indeed a space-in-between the quasi-carceral life of the camp and the ordinary life of the Italian residents, from which the migrants are entirely and deliberately excluded, as 'bodies apart' and 'in excess' (see Wong, 2015; also Agier, 2008, 2011). But the jungle also represents a first precarious entry point to the region after having crossed the Slovenian border for those who arrive via the Balkan route. These newly arrived individuals literally create makeshift camps along the riverbank using all sorts of collected materials – tents, plastic tarps, rags – while waiting to be accepted into the CARA whenever new space is made available for them. The sharp increase of arrivals in 2015, however, has also expanded the use of public space on the part of the migrants, especially in some urban green areas. This 'occupation' of public space has raised immediate concerns around issues of security and public order among the residents and the authorities. Unlike the detainees of the former CIE, the CARA 'guests' are indeed far more visible, since they are allowed to exit the centre during the day and roam freely in the surrounding areas.

> We see them more often and we consequently complain much more for their presence. The number of asylum seekers is doubled and the type of people has also changed: at present, they are found more outside the CARA than inside; I think this is due, possibly, to the growing difficulties in managing the centre, its overcrowding, the lack of alternative activities. These individuals thus prefer to spend more and more time in the city or in its immediate surroundings; they also spend lots of time along the river, and therefore the residents complain for the dirt and the mess that now characterises those areas. (L.T., present Gradisca Mayor, 31/8/2015)

Until the 'guests' were hidden inside the camp, their presence was 'tolerated' by the residents, but since their bodies began to be 'in excess' and these exotically dressed men are seen more and more often walking along the paths connecting the CARA to Gradisca or to the river, or hanging out 'in-waiting', a sense of social emergency and intolerance appears to have emerged among the local population and the authorities (see *Fain*, 2015). The 'migrants'' sudden visibility seems to represent the main problem. The fact that these individuals are mainly young men easily identifiable for their physical traits and their clothes accentuates the sense of danger they may stimulate in a local community that perceives them, literarily, as 'foreign bodies'.

Two major anchor points for the quasi-carceral geographies of these (partially) mobile subjects are both at a walking distance: an abandoned supermarket parking lot where the CARA guests play cricket and the above-mentioned Isonzo riverbank, where they literally 'camp', cook, pray, and socialise. These are buffer zones of sorts, complementary self-managed spaces. They are also the result of two major issues left unsolved by the Italian authorities: first, the shortage of hospitality facilities for skyrocketing numbers of asylum seekers (a condition getting worse month after month, especially with the recent partial blockage of the Austrian borders); second, the duration and the uncertainty of the procedure assessing their application to refugee status (see Fassin and Pandolfi, 2010). Arguably, the frequent use of former barracks and spaces of detention to host asylum seekers in dedicated camps might have played a role in the consolidation of such negative image. Indeed, that of the migrant-as-public-enemy is a trope often reinforced by the implementation of a quasi-carceral regime of mobility around these camps of forced 'care', together with the porosity of their confinement. Inside the camp, the threshold between the former area of expulsion (the former CIE) and that of hospitality (the CARA) is endlessly challenged by daily practices and the use of the CIE spaces for accommodating excessive migrants. Outside the camp, the presence of individuals freely roaming the Gradisca surroundings, and especially the sudden appearance of new makeshift camps, have created a feeling of disorder and imminent danger among the residents.

As a consequence, the 'human load' that 'exceeds' the already extra-ordinary and exclusive spatialities of the CARA breaks into a spatio-temporal regime of confinement imposed on these guests/detainees; at the same time, it highlights how the emergency security measures implemented to face the 'refugee problem' all too often seem aimed at reassuring the residents and the larger public about the perceived threat of a human wave of 'migrants' invading their everyday spaces. These security interventions are indeed often justified as a response to some alarmist comments on the part of politicians and some of the media that have long contributed to build a specifically negative image of the 'illegal' migrant as a potential public enemy. The 'human excess' made of the migrant overflow around the Gradisca centre is indeed a sort of constitutive by-product of those camp spatialities, since originated by they asylum seekers own mobility and 'non-belonging' to the social order of the territories temporarily hosting them: 'the geopolitics of migrant mobilities are produced through everyday state practices as well as by migrant strategies to move and resettle' (Ashutosh and Alison, 2012: 335). Even though these mobilities are limited to a few public spaces and buffer zones, and to particular times of the day, they are nonetheless in excess to the normal 'traffic' of people – and this paradoxical excess of mobility is what possibly creates fear in the residents as it is deemed to threaten the wider social order. What we would like to argue here is perhaps the point about excess is not so much about human excess per se but rather about an excess of human mobility/immobility. Such excess of mobility has a dual dimension: first, the mobility of the guests/detainees is in excess compared to the flows of the residents' everyday practices; second, it is an excessive mobility at the scale of the nation state and of the European continent as a whole, a mobility that seems to challenge the established territorial order since perceived as out of control. Finally, during the time spent occupying the camp, migrants and asylum seekers are enfolded within a different notion of excess – an excess of time that holds them immobile, or unable to move beyond their current situation.

* * *

When the present research terminated there was no space available to host new asylum seekers. At the same time, while the residents simply did not want to 'see' them anymore, the local authorities were debating at length on whether larger camps or minor, diffused points of hospitality – often in close contact with the local realities and the ordinary lives of the Italians – represented the best way to address the thorny challenge. The actual visibility of mobile, male, 'foreign' bodies roaming along the main roads and the fields, occupying spaces in-between 'nature' (the jungle) and the settled *civitas* of the small towns in the area, continued to provoke anxiety and concern among the residents. They consequently marked, with their own presence outside the camp, the fact that they might potentially further penetrate the ordinary social spatialities in Gradisca. The encounters between these partially (but excessively?) mobile undefined subjects and the residents seemed to be at the origin of repulsive reactions, and fierce resistance to the centre and its guests. The CARA and the makeshift encampments along the river therefore played the double function of seemingly absorbing the 'people-in-excess' and of expelling or rendering expellable those subjects with no clear and permanent role and status (Altin and Minca, 2017; also, Rahola, 2003; Bigo, 2007; Levy, 2010; Pinelli, 2015).

The ambivalent role of the camp and its quasi-carceral regime reflected the similarly ambivalent role played by the nearby border with Slovenia. The conflation of custody and restriction of mobility, on the one hand, and of assistance and hospitality, on the other, was in fact influenced by the proximity of the border and the practical consequences of the 'crime of illegal migration' in the present Italian legislation. The former Cold War border had thus become a territorial line whose crossing on the part of the migrants coming from the Balkans was filtered, moderated, and controlled by buffer zones where these same migrants were allowed to accommodate – true temporary waiting areas. These buffer zones for documentless people-in-waiting – including the 'jungle' – allowed the authorities to selectively ignore their presence and selectively admit them to the hospitality centres where their waiting spatio-temporalities would continue but in different institutional settings, as described above (Turner, 2012). Here we have identified those strategically ambivalent functions of exclusion and containment that characterise all quasi-carceral regimes. These functions 'work not only to contain mobility, but also to reconfigure and relocate national borders' (Mountz et al., 2012: 530; Malkki, 1995).

These practices and their related spatialities present themselves as a sort of enormous *rite de passage* determined by a sequence of stages where the 'migrants' were subjected, or subjected themselves, to the translation of their bodies and their individual personality into the language of their new, spatio-temporary, status of subject-in-waiting (see Gill, 2009): the procedure guiding the governance of the migrants tended to produce such ambivalence essential to the slowing down and the filtering of these growing flows of people, and to the migrants' desubjectivisation and their transformation into people in potential 'excess'. However, the porosity of such buffer zones was also appropriated by some of these individuals 'on the move': what has emerged from the interviews is the existence of networks of people – often based on ethnic belonging and area of provenience – following specific routes and incorporating these ambivalences as a strategy. For example, among some networks of individuals coming from Pakistan and Afghanistan, this border region is considered a privileged one because, compared to others, 'there, they treat you well'. What we are trying to say here is that the ambivalence, the holes-in-the-net, the porosity of the controls and the inefficiencies of the hospitality provided by the camp were strategically used by some of these asylum seekers to produce new forms of subjectivities based on the quasi-carceral mobilities in interstitial spaces like the jungle and the riverbank: strategies and tactics, which became forms of resistance using precisely the ambivalence of the 'refugees' management system' (Ashutosh and Alison, 2012). Despite being regularly subjected to biopolitical techniques of governmentality, the 'asylum seekers claimed the right to question how they were governed and thus constituted themselves not merely as governed individuals but also as political subjects' (Conlon, 2013: 145).

For the 'migrants', to act and to voice their concerns by becoming present and visible in these spaces meant being able to reject their objective vulnerability as typical of the condition of nameless victims, despite this form of resistance taking place in the same spaces that are at the foundation of such vulnerability, that is, the detention and hospitality centres: more than bare life, they embodied 'life that resists' (Agier, 2008). If it is true, as noted by Foucault (1975; see also Moran, 2015), that the carceral regime is founded on the production of disciplined and docile subjects, it is thus perhaps useful to recall how the carceral apparatus is in fact not limited to functions of detention, custody, and control, but it is eminently, and perhaps first of all, a discursive formation that guides the behaviour and produces the very horizon of possibilities of the subjects through which these same subjects are defined and often end up defining themselves. Such ambivalence of the camp, together with the production of subjects and 'bodies in excess', seems to be the fundamental source and the arcanum of all politics aimed at expanding the production and reproduction of carceral and quasi-carceral mobilities, inside and outside the camp.

References

Il Piccolo (2016) Gorizia, 21 indagati per la vicenda: Connecting People, 8 February 2016.

Agier M (2008) *On the Margin of the World*. Cambridge: Polity Press.

Agier M (2011) *Managing the Undesirables*. Cambridge: Polity Press.

Altin R and Minca C (2017) Exopolis reloaded: Fragmented landscapes and no man's lands in the Italian North-East. *Landscape Research*.

Armstrong S (2015) The cell and the corridor: Imprisonment as waiting, and waiting as mobile. *Time & Society* 00: 1–22. Available at: http://eprints.gla.ac.uk/106124/1/106124.pdf

Ashutosh I and Alison M (2012) The geopolitics of migrant mobility: Tracing state relations through refugee claims, boats, and discourses. *Geopolitics* 17(2): 335–354.

Bigo D (2007) Exception et ban: A propos de l'Etat d'exception. *Erytheis* 2: 115–145.

Conlon D (2011) Waiting: Feminist perspectives on the spacings/timings of migrant (im) mobility. *Gender, Place & Culture* 18(3): 353–360.

Conlon D (2013) Hungering for freedom: Asylum seekers' hunger strikes – rethinking resistance as counter-conduct. In: Moran D, Gill N and Conlon D (eds) *Carceral Spaces: Mobility and Agency in Imprisonment and Migrant Detention*. Farnham: Ashgate, 133–148.

De Genova N and Peutz N (eds) (2010) *The Deportation Regime*. Durham, NC: Duke University Press.

Fain F (2015) Torrenti: 'Cara di Gradisca affollato? La soluzione è temporanea', *Il Piccolo*, 20 October 2015.

Fassin D and Pandolfi M (eds) (2010) *Contemporary States of Emergency*. New York, NY: Zone Books.

Felder M., Minca C. and Ong CE (2014) Governing refugee space: The quasi-carceral regime of Amsterdam's Lloyd Hotel, a German-Jewish refugee camp in the prelude to World-War-Two. *Geographica Helvetica* 69: 365–375.

Foucault M (1975) *Surveiller et punir. Naissans de la prison*. Paris: Gallimard.

Gatti C (2014) 'Mafia romana', il business dei centri di accoglienza. *Il Sole 24 Ore*, 12 March 2014. Available at: www.ilsole24ore.com/art/notizie/2014-12-03/mafia-romana-business-centri-accoglienza--121834.shtml?uuid=ABNP7OLC

Gill N (2009) Governmental mobility: The power effects of the movement of detained asylum seekers around Britain's detention estate. *Political Geography* 28(3): 186–196.

Levy C (2010) Refugees, Europe, camps/state of exception: 'Into the zone', the European Union and extraterritorial processing of migrants, refugees, and asylum-seekers (theories and practice). *Refugees Studies Quarterly* 29(1): 91–119.

Majcher I (2013) *'Crimmigration' in the European Union through the lens of immigration detention*. Global Detention Project, Working Paper 6.

Malkki L (1995) Refugees and exile: From 'refugees studies' to the national order of things. *Annual Review of Anthropology* 24: 495–523.

MEDU (Medici per i Diritti Umani) (2013) *Arcipelago CIE*. Formigine: Infinito.

Minca C (2005) The return of the camp. *Progress in Human Geography* 29(4): 405–412.

Minca C (2015a) The biopolitical imperative. In: Agnew J, Mamadouh V, Secor AJ and Sharp J (eds) *The Wiley-Blackwell Companion to Political Geography*. London: Wiley-Blackwell, 165–186.

Minca C (2015b) Geographies of the camp. *Political Geography* 49: 74–83.

Moran D (2012) 'Doing time' in carceral space: TimeSpace and carceral geography. *Geografiska Annaler: Series B, Human Geography* 94(4): 305–316.

Moran D (2015) *Carceral Geography: Spaces and Practices of Incarceration*. Farnham: Ashgate.

Moran D, Gill N and Conlon D (eds) (2013) *Carceral Spaces: Mobility and Agency in Imprisonment and Migrant Detention*. Farnham: Ashgate.

Mountz A (2011) The enforcement archipelago: Detention, haunting, and asylum on islands. *Political Geography* 30: 118–128.

Mountz A, Coddington K, Catania RT and Loyd JM (2012) Conceptualizing detention, mobility, containment, bordering and exclusion. *Progress in Human Geography* 37(4): 522–541.

Ong A (2003) *Buddha Is Hiding: Refugees, Citizenship, the New America*. Berkeley, CA: University of California Press.

Peters K (2014) Tracking (im)mobilities at sea: Ships, boats and surveillance strategies. *Mobilities* 9(3): 414–431.

Pinelli B (2015) After the landing: Moral control and surveillance in Italy's asylum seeker camps. *Anthropology Today* 31(2): 12–14.

Rahola F (2003) *Zone definitivamente temporanee*. Verona: Ombre Corte.

Rivoli A (2014) Immigrazione e dispositivi di esclusione: Il caso del CIE-CARA di Gradisca. Unpublished masters thesis, University of Trieste.

Sigona N (2014) The politics of refugee voices. In: Fiddian-Qasmiyeh E, Loescher G, Long K and Sigona N (eds) *The Oxford Handbook of Refugee and Forced Migration Studies*, Oxford: Oxford University Press, 369–382.

Sorgoni B (ed) (2011) *Etnografia dell'accoglienza*. Rome: CISU.

Turner J (2012) Criminals with 'community spirit': Practising citizenship in the hidden world of the prison. *Space and Polity* 16(3): 321–324.

Turner J (2013a) Disciplinary engagement with the prisons, prisoners and the penal system. *Geography Compass* 7(1): 35–45.

Turner J (2013b) Re-'homing' the ex-offender: Constructing a 'prisoner dyspora'. *Area* 45(4): 485–492.

Verdirame G and Harrel Bond B (2005) *Rights in Exile*. New York, NY: Berghahn.

Vicedomini D (2016) La truffa dell'accoglienza, anche Vip tra gli indagati: Intascavano i soldi pro migranti e traevano vantaggi fiscali. *Il Messaggero Veneto*, 9 February 2016.

Wong TK (2015) *Right, Deportation and Detention in the Age of Immigration Control*. Stanford, CA: Stanford University Press.

4 'Unruly mobilities' in the tracking of young offenders and criminality

Understanding diversionary programs as carceral space

Elaine Fishwick and Michael Wearing

The aim of this chapter is to understand the specifics of governing young offenders and their criminality via diversionary programmes, in the context of the Australian state of New South Wales (NSW). In doing so, we bring a geographical and carceral mobilities focus to the study of youth justice. NSW has developed a suite of diversionary measures, that is, pre-sentence and post-sentence options to keep young people *away* from the formal court system. However, whilst we argue that this presents a clear youth justice policy (*Young Offenders Act NSW 1997*), these diversionary measures manifest as liminal carceral spaces where young people are governed and where detention in an institution is positioned on a continuum of punitive sanctions. Such diversionary programmes are contextualised in newly articulated relations of surveillance and security in the antiterror state. They categorise criminality and demonise the non-White other (Monaqhan, 2012) in tracking the movement of youth deemed 'at risk' of crime in non-penal spaces of the community (Follis, 2015). Indeed, we also propose that in a neocolonial state such as NSW, contemporary articulations of hyperincarceration, emerge from landscapes of colonial carceral regimes, whilst also embodying distinct characteristics and patterns of resistance.

In this chapter we explore the extension of the carceral state into the community and into the lives of young offenders where the legal-administrative measures of diversionary programmes create long distance control and discipline (including self-discipline), and punishment over a broad geography. We define the movement encompassed within diversionary measures as 'carceral mobilities' – as disciplinary mobility or the rational and technical means of governmental control over youth offender space within the community orientation of NSW juvenile justice (Brown, 2014; Moran et al., 2012). Such measures can coerce, constrain, and potentially immobilise the lived time and spatial freedoms of such youth creating liminal carceral spaces in the home, street, or shopping mall.

The need to govern the behaviours of young offenders is driven by their 'unruly mobilities'. Young offenders are often seen to be unpredictable and ungovernable, ill-disciplined and poorly socialised (Moran et al., 2012). The criminalisation of 'unruly mobilities' has hence produced new spaces and practices of youth justice in New South Wales. Patterns of mobility and movements are orchestrated and mapped by specific mobile technologies and practices: for example, curfews, behaviour management, and probationary regimes. These are articulated in youth justice policy discourse and in the broader practices that govern youth freedoms. These include the policing of young people in public and private spaces; the labelling of 'at risk' youth; restrictions on movement and association in private and public spaces; the de facto criminalisation of 'sexting'; the disciplinary technologies of youth bail and of diversionary practices themselves (Fishwick, 2007, 2015).

We begin our discussion by exploring the constitution of youth as potentially 'unruly' and disruptive populations and map out the dimensions of carceral regimes in relation to criminalised young people. We then contextualise our discussion by arguing that in neo-colonial and postcolonial societies like Australia, carceral regimes have distinct features where First Nations[1] peoples were and continue to be governed differently from settler populations (see also Coddington, this volume). The chapter then explores the reach of diversionary programmes in to the lives of individual young people and crime prevention strategies more generally, before examining the capacity of geographical communities to break the cycle and churn of hyperincarceration by instituting community-based justice reinvestment strategies.

Unruly mobilities of youth

Conventionally those at risk who are vulnerable, marginalised, and usually highly disadvantaged are categorised as 'neglected children', 'truants', 'potential car thieves', 'taggers', or 'shoplifters', and are identified and tracked and their details shared between police, welfare, health and education authorities. These labelling activities do not necessarily orchestrate 'moral panics' but are more likely to enable identifiable moral clusters of known youth 'suspects' based on their whereabouts in geographic domains. Such known 'clusters' of youth are seen to engage in 'unruly' mobilities in the spaces shared with other citizens (such as shopping malls or parks), their identities circulating amongst experts and front-line police and welfare agents. Effectively these new 'regimes of circulation' (Follis, 2015) are formed through renewed efforts by authorities to step up police monitoring and penal-welfare surveillance (Philo, 2014).

Figure 4.1 indicates how we map out conceptually the spaces and practices discussed in relation to recent reforms to the management of youth and crime in New South Wales. We have used the carceral turn in geography (Moran, 2015; Philo, 2014) and combined this with our own focus from within a critical criminology approach to highlight four connected dimensions – 'regimes' through to 'clusters', 'networks', and 'practices' – which are part of a dynamic feedback loop that illustrates the 'churning' powers of social control that are exercised as part of diversionary programmes. These powers work in the spheres of the public and private, micro and macro, and from incarceration in imprisonment spaces to non-incarceration in the community and the home. The arrow in the middle of the diagram between perceived and actual freedoms and confinement indicates liminal carceral spaces in between these poles.

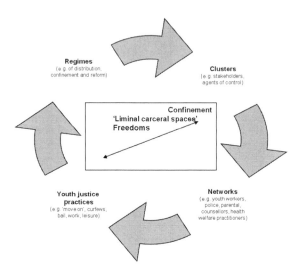

Figure 4.1 Mapping the practices and spaces of carceral
mobilities. (Designed by the authors.)

We also use Moran's (2013, 2015) helpful term 'liminal carceral space' (Figure 4.1) throughout to portray how mobility is fundamental to regimes of control beyond spaces of physical and territorial incarceration. The liminal space that youth occupy – restricted yet not incarcerated – create modes of psycho-emotional constraint (e.g., paranoia) more than physical constraints, indicative of Foucault's (1977) take on modes of surveillance in the modern era.

We have briefly sketched some of the key definitions of Figure 4.1 noting the ever-changing, self-disciplinary, and fluid basis of youth offender identity within diversionary programmes that govern youth mobilities. Next, we consider the context for the governance of unruly youth mobilities through the (im)mobilising nature of diversionary programmes. Accordingly, we examine the colonial relations that have long shaped the governance of youth offenders.

The geography of penal colonial relations

The founding of a penal colony in NSW by the British in 1788 was an act of colonial expansion in a search for new economic resources. White colonial settlement and colonisation of Australia resulted as a partial solution to the rapid growth in prison numbers in the UK with the transportation of adults and children to a distant carceral space (Hughes, 1987; Muncie, 2009). This gateway event led to a cataclysmic disruption to the world of the original custodians of the lands we now know as Australia. Despite the fact that social, juridical and economic welfare systems of the new colony incorporated many traditions of the colonising power, they were distinct and were fashioned by colonial and convict narratives and penal practices that led to a very different society. It was a place where inhabitants of New South Wales were living in militarised 'landscapes of exile' (Cunneen et al., 2013: 21). These neocolonial conditions of emergence of NSW continue to permeate contemporary penal culture, carceral practices and discourses and their enduring legacy can be found in the contemporary 'hyper-incarceration' of Indigenous people (Cunneen et al., 2013: 20). Colonial practices were partly justified by the eugenicist belief that 'Aboriginal' peoples were not fully developed humans. Subsequently, they were first exiled from their lands and placed on segregated reserves and missions (Hogg, 2001). These practices fractured the custodianship of country and severely disrupted kinship, social, cultural, and traditional juridical relations (Behrendt et al., 2009; Blagg, 2008; Hogg, 2001).

From the late eighteenth century onwards the police in the new town of Sydney were militarised and were not only engaged in the control of crime, but were involved in frontier expansion and in the moral and behavioural regulation of the state's inhabitants. The police were responsible for regulating health and sanitation, controlling vagrancy, licensing, registering people for elections, care and protection, establishing local government, and a whole host of broader administrative activities that were needed in building a new social infrastructure (Hogg and Golder, 1987). These histories can also be reframed as sites for the creation of discipline over mobilities within the community and as antecedents to diversionary and community orientations in NSW youth justice. Colonial Australia provided many of the bio-disciplinary themes and archetypes within prison buildings worldwide such as panoptic surveillance. This discipline of convicts and Aboriginal and Torres Strait Islander peoples can be partially understood in Foucauldian analysis as a transformational shift from physical punishments to psychological and emotional punishments that involved modes of disciplining the mobilities of convicts, for example, through indentured labour in the late eighteenth and early nineteenth centuries in the West (Foucault, 1977).

As transportation declined towards the middle of the nineteenth century, convicts were freed and voluntary emigration to Australia began. However, Aboriginal and Torres Strait Islander reserves and missions continued to be key sites of colonial penality. According to Hogg (2001) carceral practices like physical punishment, rationing, segregation, and forced removal were fundamental to colonial societies and continued to permeate the treatment of Aboriginal and Torres Strait Islander people in institutional and non-institutional settings, in ways that were no longer tolerated in non-colonial societies. For example, public executions, floggings, and restraint by chains were used against Indigenous people long after they had ended for white populations, and were justified on the basis that Aboriginal people were not capable of the self-disciplinary techniques of contemplation and rehabilitation that underpinned imprisonment or probation and parole (Cunneen et al., 2013; Hogg, 2001).

The carceral regimes of the penal colony extended beyond the prison to the geographically isolated missions and reserves of displaced communities. Here, colonial carceral practices also encompassed the day-to-day administration of missions and reserves where Aboriginal and Torres Strait Islander people were relocated. Individuals were not free to leave the reserve or to marry, nor were they allowed to engage in ceremony, culture, and language, or to seek employment without permission. Families and individuals were provided with rations and were often not paid for work. They also faced further exile, physical punishment, or imprisonment if they disobeyed the police and welfare authorities.

Additionally, under the pretext of care and protection, children could be taken away from families and placed in institutions, on farms or in homes with white families – becoming a source of indentured labour (see Behrendt et al., 2009; Cunneen et al., 2013; Hogg, 2001). Children removed from their families came to be known as the *Stolen Generations* (Cunneen et al., 2013; HREOC, 1997). This was done not only to inculcate what the *Bringing Them Home Report* has called 'the diligent subservience of thought desirable in servants and the working class' (HREOC, 1997: 33), but also to eliminate First Australians' cultural practices and kinship networks. Girls and young women were morally condemned as sexually promiscuous and as bad mothers and were subject to intense disciplinary regulation, punishment and exploitation (Carrington and Pereira, 2009).

The removal of children from their families was a practice that continued long into the twentieth century (Cunneen et al., 2013; HREOC, 1997). Consequently, the impacts of these policies are still felt today. The earlier forms of 'penal-welfarism' legitimated in the era of the stolen generations are today inscribed in the day-to-day management and long distance policy controls and evinced in high rates of child abuse, incarceration, deaths in custody, and criminalisation of Aboriginal youth (Carrington and Pereira, 2009: 104). To some extent Aboriginal and Torres Strait Islander peoples continue to be constituted an 'unruly' mobile population and are policed today within the shadow of these colonial and neocolonial developments in NSW and around Australia.

Consequently, we see mutually reinforcing cycles of disadvantage and criminalisation in the numbers of Aboriginal and Torres Strait Islander young people flowing between care and protection systems, juvenile detention and the adult gaol. There is an acute overrepresentation of Aboriginal and Torres Strait Islander children and young peoples in key indicators of poverty, social deprivation, care and protection notifications and in their contact with the criminal justice system. Children continue to be removed and placed into wardship disproportionally, in far higher numbers than non-Indigenous children, and they are often placed in institutions far away from their traditional lands and communities (AIFS, 2014; AIHW, 2012; Carrington and Pereira, 2009). A recent report by Australia's Social Justice Commissioner found that the Aboriginal reimprisonment rate, which is 58 per cent within a ten-year period, is actually higher than the Aboriginal school retention rate for the high school years from ages 12 to 18, currently running at 46.5 per cent (AHRC, 2014: 100). Nationally about half of the young people in detention are Indigenous, and on an average night 52 per cent of young people in detention were Aboriginal, and the rate of Aboriginal youth in detention is 25 times that of non-Aboriginal youth (AHRC, 2014).

Governments continue to intervene in Aboriginal and Torres Strait Islander people's lives, especially in rural and remote communities, in ways that would not be contemplated in non-Indigenous communities. In the Northern Territory, for example, the Commonwealth government introduced a series of policies known collectively as the Northern Territory Intervention, placing military personnel temporarily in outback communities to administer public policy, and tightly regulating income expenditure: with the introduction of welfare cards that control where people can spend; the introduction of welfare penalties for families of children missing school; and replacing locally controlled governance mechanisms with direct government management (Cunneen et al., 2013; see also Coddington, this volume).

Diversionary measures can be understood as a further mechanism by which the state – and carceral regimes – move beyond traditional spaces of detention and confinement and into everyday space. These measures work specifically to govern the continued conceptualisation that specific societal groups are 'unruly'. In what follows we attend to the liminal spaces such programmes produce as they govern youth movement. However, as we shall also discuss, Indigenous communities continue to negotiate policy space in a way that aims to disrupt these established cycles of disadvantage and hyperincarceration, endeavouring to keep children and young people in the community and away from places of detention altogether.

Mobile liminal confinements through diversion

One of the many strategies utilised in youth justice for deterring crime and preventing repeat offending is that of diversion, both away from the court and away from detention and into the community (Cunneen et al., 2013; Muncie, 2009). Diversion is accepted as a key principle underpinning children's rights worldwide and is articulated in the UN Convention on the Rights of the Child (Article 37), the Beijing Rules, the Riyadh Guidelines and many regional human rights instruments (Weber et al., 2014). Yet, as we argue here, the impact of diversion is not always benign, and indeed diversion and incarceration are inextricably linked. As discussed earlier the geographical turn in understanding incarceration as a mobile and liminal condition is central to how we understand current discourses of diversion and prevention in the NSW system related to youth justice. Here we have reframed such practices as part of the extension of the carceral state through disciplinary mobility (Moran et al., 2012).

Techniques of incarceration that manifest in the local community and in wider society are now possible in unprecedented ways that were not feasible in earlier periods due to lack of infrastructure and technology for security; such as DNA analysis; high-powered satellite surveillance cameras; Taser guns; electronic tracking devices; Internet surveillance; and new security forces such as the recent Australia-named 'Border Force' introduced to police migration and 'terrorism'. As Brown has argued, 'a geographically expansive understanding of the carceral opens up incarceration as a multi-institutional, fluid, and indeterminate practice' (2014: 386). Hyperincarceration not only enables constraint within penal and remand institutions but also governs socio-spatial movements outside of such places of confinement.

In NSW, diversionary programmes are generally targeted at less serious offences and first time offenders, and include diversion away from the criminal justice system, from the court, or from custodial detention (see the *Young Offenders Act 1997*

New South Wales). As such, these programs govern and *direct* mobilities – that is, where young people go (or do not go). Programmes include a suite of warnings cautions and fines administered by both the police and the court; and forms of youth justice conferencing that bring all parties – perpetrators and victims and sometimes family members or others – together to limit young people's further entry into the formal criminal justice system by developing a mutually agreed outcome plan. Youth diversionary programs can also include referral to a range of health and drug and alcohol programs such as the 'Youth on Track' program aimed to keep those at risk of reoffending over the age of 10 from re-entering the criminal justice system through case management (see for a current overview: AIHW, 2016; Fishwick, 2015). [2] Each of these involve different degrees of regulation of young people's behaviour, movement, and time. Accumulated empirical evidence demonstrates that access to diversionary options is heavily racialised and far fewer Indigenous young people enter diversionary programs in comparison to their non-Indigenous counterparts (Cunneen et al., 2013).[3]

Cautions, community service orders and restorative justice outcome plans that are part of diversionary programmes often include conditions that require compliance by young people, or there is a risk of further sanction. Many points of compliance govern young people's movements and their ability to meet with friends and acquaintances. Courts hold the power to impose an order restricting whom young people can meet with and where they can go. If a young person breaches the order they may be liable for a sentence of imprisonment (see *Justice Legislation Amendment [Non-Association and Place Restriction] Act 2001 [NSW]*). Placing a curfew on an individual's presence in public after certain hours has been an option used in probation, community service orders, and sentencing determinations for many years. Curfews are commonly used as a condition of bail (see below for further discussion) and provide police with the power to do curfew checks at young people's homes. These can include the police coming to a house in the middle of the night and demanding to see if a young person is there, checking on who a young person is with; or waiting at train stations or other transport hubs to make sure that they are travelling back on time (NSWLRC, 2012; Shopfront Youth Legal Centre, 2015a, 2015b; Wong et el., 2010). Technical breaches of curfews have meant that young people have been arrested and placed in detention as a result of minor breaches of such conditions (NSWLRC, 2012; Wong et al., 2010). In turn, diversionary programmes *create* unruly mobilities as the carceral state extends into the spaces of everyday life.

Other features of the governance of youth relate to bail. In NSW the police, the court or a judicial officer can grant or deny a young person bail in relation to their offence (*Bail Act 2013*). Utilising a risk matrix, bail can either be granted or if it is denied the young person can be remanded into custody to wait for their court appearance. Clearly this measure both governs 'unruly' mobilities, whilst also creating a series of subsequent (im)mobilities on a case by case basis. Indeed, the police or the court can attach restrictions/conditions to bail relating to movement. These include mandatory reporting to police or juvenile justice authorities, governance of who the young person can be with, where they can live, and their attendance at school, college, job training, health programmes, and so on. The largest proportion of young people in custody in NSW and around Australia continues to be young people who are placed there on remand (that is, they are there without having been convicted of any offence and are awaiting sentence and numbers are increasing (DJJ, 2014; Stubbs, 2010; Weatherburn and Fitzgerald, 2015). Accordingly, bail occupies a liminal space where mobilities associated with release, or even those associated with diversionary discipline are put 'on hold' until a court decision is made.

Being granted bail or being remanded in custody are examples of liminal spaces and conditions where the tight regulation of mobility occurs both inside and outside of the institution; and where the boundary between the prison and the community is porous and inconstant. Yet it remains that the impact of being placed on remand is so much more profound as young people are separated from community, family and from school. Research evidence consistently shows that young people on remand receive tougher sentences than those on bail and that the chances of criminal recidivism are increased (NSWLRC, 2012).

Diversionary practices such as those outlined above help to illustrate the extension of the state into the lives of young people, beyond traditional spaces of incarceration. Diversionary measures have been described in the criminological literature as 'pre-emptive', that is, they exemplify a shift from prevention of crime to pre-emption of crime and future risk (Zedner, 2007). This shift is indicative of a new set of social ordering practices and, we would argue, creates new liminal carceral spaces and new forms of mobilities that engage communities, volunteers, and individuals. We now turn our attention to understanding youth as unruly and criminogenic to explore how they are governed in public and private spaces, and how these governance measures involve a range of interventions in young peoples' lives.

Crime prevention and the policing of unruly mobilities in public and private spaces

Young people are conceptualised and stereotyped in public space (and increasingly private space) as potentially disruptive and threatening to other 'legitimate' users of the space in question. In NSW, young people's presence in, and access into and out of streets, parks and other public areas has been regulated and criminalised in a number of ways, including public order legislation such as the *Summary Offences Act NSW 1988*. This latter measure was introduced, ostensibly, as an 'anti-gang' measure (although there was no empirical evidence of US-style gangs). Police were given the powers to order young people in groups of three or more to disperse; to 'move on' *before* they become 'disruptive'. These pre-emptive, dispersal powers to move young people from one place to another illustrates the disciplining of mobilities created by the threat of 'unruliness'.

Concerns with governmental control of public space have been intensified in the wake of 9/11 when Australian governments, like others around the world, increased security measures including the monitoring and surveillance of public space, of protest, and of 'troublesome' populations. After a series of street disturbances in Sydney and in rural towns, police powers in NSW were extended and consolidated (see *Law Enforcement [Powers and Responsibilities] Act 2002 and subsequent amendments in 2002, 2005 and 2007)*. Accordingly the police now have the authority to direct young people to leave public places if they are causing an obstruction; are deemed to be engaged in harassment or *might* be engaged in harassment; are intoxicated by drugs or alcohol, or suspected of supplying drugs; or are 'causing or likely to cause fear to another person ...' (Shopfront Legal Centre, 2011: no page). The police also have emergency powers to break up groups (or any members of that group) within a designated 'target area' (see *Law Enforcement [Powers and Responsibilities] Act* s87MA 2005). And, according to Shopfront Legal Centre, 'it appears that the police can do this even if they have no reason to think that breaking up the group will help prevent or control the public disorder' (Shopfront Youth Legal Centre, 2011: no page). Police also have the power to control the movement of cars in and out of suburbs, to search cars and to lock down streets and suburbs in designated 'target areas' (Loughnan, 2009; Lynch et al., 2010).

In NSW Aboriginal young people are more often than not subject to far heavier surveillance and policing interventions than other groups (Cunneen et al., 2013). In parts of the city and in towns where there are high concentrations of Aboriginal youth, local councils and police have controlled the presence of Aboriginal youth on the streets with the targeted street sweeping powers, frequent stop and search, and night time curfews for children and teenagers (Sanders and Grainger, 2003). Moreover, with the privatisation of public space – especially retail space, and the growth of shopping malls (Flint, 2006; White, 1990) – the control of young people's movements into and out of space have been subject to private policing by security guards (see also McLean, 2005). In Australia, the enclosed purpose-built shopping mall has become the central meeting, shopping and business point of many suburbs and rural towns. Young people are photographed and filmed by staff and banning orders are enforced using this identification. On occasions, young people are actively discouraged from gathering near shops by the broadcasting of classical music, although the Mosquito – the high-pitch youth deterrent noise used in the UK to disperse young people – has not been used to date (UK Children's Commissioner, no date). Security guards have the right to remove people by reasonable force from private property, and banning orders are imposed at the discretion of the owners or the managers of private property.

In rural areas in Australia essential services such as post, welfare, and employment services, and health centres are located in privatised spaces and if excluded, young people's rights to freedom of movement are not only denied but they are prohibited from exercising their social and economic citizenship rights (Clancey et al., 2002; Grant, 2000). This illustrates the stark restriction of youth mobilities in spaces of supposed 'freedom', where such spaces become liminal – neither free nor wholly confining – but where unruliness (or perceived unruliness) is enough for governance measures to be enacted.

Together, these examples illustrate how challenges to 'law and order' in local environments elicit extra measures to govern youth. Commonly, as indicated in Figure 4.1, clusters of localised stakeholders (including police, parents, neighbours, local councillors, social and youth workers, private retail managers) are complicit in such measures of confinement, seeing it as part of their responsibility to manage and take responsibility for the behaviour of young people. Such social management, we argue, has transformed the management of youth mobilities, resulting in practices of incarceration spreading into shopping malls, streets, parks and to other leisure activities.

The geography of resistance: Justice reinvestment

As noted, recent research in Australia on justice reinvestment has stressed how the geographies of imprisonment across the country intersect with those of poverty and Aboriginality (Brown et al., 2015). As in the literature on hyperincarceration and people of colour in the US and Europe, and as discussed above, the cycle of criminalisation and institutionalisation for many young Aboriginal people is a regular feature of growing up (Cunneen et al., 2013; Brown et al., 2015). However, over recent years, in contrast to those justice reinvestment programs and policies in the US that appear to extend carceral practices of the prison into the community without welfare provision, local Aboriginal communities in areas like Redfern in the city of Sydney ('Clean Slate without Prejudice' program) and Bourke, a rural town in the West of the state (Maranguka Justice Reinvestment), have, through consultation with the local community, the local council, the police, and other key agencies, developed holistic, community-based strategies that are informed by principles of self-determination, cultural safety and cultural health. They are community development responses to marginalisation and criminalisation that aim to improve the health and well-being of the geographical and social community by being managed by local people in partnership with other agencies. Although only at an early stage, there are already marked reductions in offending and re-offending, and a greater sense of community well-being and cultural safety in both locations (see AHRC 2014; Amnesty International, 2015; Brown et al., 2015).

Other Indigenous communities concerned about young people being at risk have also introduced a range of diversionary and community policing options to ensure that young people who are out at night, are safe. The Report of the Australian Social Justice Commissioner provides details of the range of voluntary and diversionary programs developed to offer alternatives to being out on the streets, including Aboriginal-run community transport schemes introduced to help young people to get home or to safe places within the community (AHRC, 2014). These programs offer some potential for autonomy and freedom of movement for Aboriginal youth away from governance by the state, invoking cultural and community-based disciplinary mobility. The focus here is not simply on individual freedoms, autonomy, and individual desistance from crime, but also on sovereign challenges to colonial carceral regimes. The emphasis is on strategies developed from the ground up, where community elders direct program development rather than government and non-government organisations and services. The diversionary and crime prevention programs are viewed holistically as part of broader cultural safety and health programs, where individuals are viewed as members of traditional kinship and community networks. Despite localised success and the potential of justice reinvestment, the realm of choices and decisions available to diverted offenders remains, in reality, quite limited and constrained by the broader neocolonial relations of other realms of government and the reluctance of the wider Australian population to accept self-determined Indigenous response to injustice.

Conclusions

This chapter has considered new regimes of youth justice within NSW and described how such programmes are articulated within the landscape of neocolonial penal carceral regimes. It has also provided a broader understanding of such shifts and trends in diversionary practices, and the circulation of 'prevention' discourse and select measures to control young people. Notably, this chapter has argued that such regimes hinge on mobilities – the perceived need to govern 'unruly' mobilities through the control of how, where and when youth can and cannot move. The geographical turn to carceral mobilities has enabled a deeper socio-spatial understanding of how everyday spaces are governed, creating 'liminal' spaces of semi-confinement. Moran's (2015) work on liminal carcerality as 'seeming imprisonment' is helpful in describing such practices especially in non-statutory policy regulation and discretionary police diversions. This chapter has thus outlined how current NSW diversionary programs illustrate the governing of 'unruly mobilities' in the community and in the home, as an extension of the carceral state in NSW youth justice reform. These measures primarily use a risk-filled discourse to govern movement, space, and mobilities for diverted youth offenders. Yet, as the chapter also indicates, there is promise with spaces for transformative locally based community actions to provide protection from the worst excesses of such disciplinary mobility.

Notes

1 We use the term 'First Nations' people' to refer to those original custodians of land around the world displaced by colonial powers. In reference to Australia we use the following terms: 'First Australians', 'Indigenous peoples', and 'Aboriginal and Torres Strait Islander peoples'. The terms vary in response to the source materials used.
2 These also include Traffic Offender Programs, MERIT (Magistrates' Early Referral Into Treatment), Rural Alcohol Diversion, the Adult Drug Court, the Youth Drug and Alcohol Court, and Community Conferencing for Young Adults.
3 When similar age, gender, offence type, and previous offence history are taken into consideration.

References

AHRC (Australian Human Rights Commission) (2014) *Social Justice and Native Title Report 2014*. Aboriginal and Torres Strait Islander Social Justice Commissioner. Available at: www.humanrights.gov.au/social-justice-and-native-title-reports

AIFS (Australian Institute for Family Studies) (2014) *Child Protection and Aboriginal and Torres Strait Islander Children*. Available at: aifs.gov.au/cfca/publications/child-protection-and-aboriginal-and-torres-strait-islander-children

AIHW (Australian Institute for Health and Welfare) (2012) *Children and Young People at Risk of Social Exclusion: The Links Between Homelessness, Child Protection and Juvenile Justice*. Canberra, ACT. Available at: www.aihw.gov.au/WorkArea/DownloadAsset.aspx?id=60129542238

AIHW (Australian Institute for Health and Welfare) (2016) *Youth Justice Supervision in New South Wales*. Available at: www.aihw.gov.au/youth-justice/states-territories/nsw/

Amnesty International (2015) *A Brighter Tomorrow: Keeping Aboriginal Kids in the Community and out of Detention in Australia*. Available at: www.amnesty.org.au/images/uploads/aus/A_brighter_future_National_report.pdf

Behrendt L, Cunneen C and Libesman T (2009) *Indigenous Legal Relations in Australia*. Oxford: Oxford University Press.

Blagg H (2008) *Crime, Aboriginality and the Decolonization of Justice*. Sydney: Hawkins Press.

Brown E (2014) Expanding carceral geographies: Challenging mass incarceration and creating a 'community orientation' towards juvenile delinquency. *Geographica Helevetica* 69: 377–388.

Brown D, Cunneen C, Schwarz M, Stubbs J and Young C (2015) *Justice Reinvestment: Winding Back Imprisonment*. London: Palgrave Macmillan.

Carrington K and Pereira M (2009) *Offending Youth: Sex, Crime and Justice*. Annandale, Sydney: The Federation Press.

Clancey G, Doran S and Robertson D (2002) *Creating the Space for a Dialogue: A Guide to Developing a Local Youth Shopping Protocol*. NSW: Shopping Centre Protocol Project.

Clean Slate Without Prejudice www.healthinfonet.ecu.edu.au/key-resources/programs-projects?pid=1263

Cunneen C, Baldry E, Brown D, Brown M, Schwartz M and Steel A (2013) *Penal culture and Hyper Incarceration*. Farnham: Ashgate.

DJJ (NSW Department of Juvenile Justice) (2014) *DJJ Statistical Information Systems*. Available at: www.djj.nsw.gov.au/statistics.htm

Fishwick E (2007) Sharing stories about labour: Youth justice strategies in New South Wales and the UK. In: Roberts R and McMahon W (eds) *Social Justice and Criminal Justice*. London: Centre for Crime and Justice Studies King's College London, 442–661.

Fishwick E (2015) 'When the stars align': Policy decision making in the NSW Juvenile Justice System 1990–2005. Unpublished doctoral dissertation, Sydney University.

Flint J (2006) Surveillance and exclusion practices in the governance of access to shopping centres on periphery estates in the UK. *Surveillance & Society* 4(1/2): 52–68.

Follis L (2015) Power in motion: Tracking time, space and movement in the British Penal Estate. *Environment and Planning D: Society and Space* 33(5): 945–962.

Foucault M (1977) *Discipline and Punish: The Birth of the Prison*. London: Penguin.

Grant C (2000) Banning the banning notice: Banning Notices, shopping centres and young people. *Alternative Law Journal* 25(7): 32–34.

Hogg R (2001) Penality and modes of regulating indigenous people in Australia. *Punishment and Society* 3(3): 355–379.

Hogg R and Golder H (1987) Policing Sydney in the late nineteenth century. In: Finnane M (ed) *Policing in Australia: Historical Perspectives*. Sydney: University of New South Wales Press, 59–73.

HREOC (1997) *Bringing Them Home: A Report of the National Inquiry into the Separation of Aboriginal and Torres Strait Islander Children from Their Families*. Sydney: Commonwealth of Australia. Available at: www.humanrights.gov.au/sites/default/files/content/pdf/social_justice/bringing_them_home_report.pdf

Hughes R (1987) *The Fatal Shore*. London: Vintage.

Loughnan A (2009) The legislation we had to have?: The Crimes (Criminal Organisation Control) Act 2009 (NSW). *Current Issues in Criminal Justice* 20(3): 457–565.

Lynch A, McGarrity N and Williams G (2010) *Counter Terrorism and Beyond: The Culture of Crime and Justice after 9/11*. Abingdon: Routledge.

Maranguka Justice Reinvestment www.healthinfonet.ecu.edu.au/key-resources/programs-projects?pid=2586

McLean G (2005) In the hood. *The Guardian Online*, 14 May. Available at: www.theguardian.com/politics/2005/may/13/fashion.fashionandstyle

Monaqhan J (2012) Terror carceralism, surveillance, security, governance and de/civilization. *Punishment and Society* 15(1): 3–22.

Moran D (2013) Carceral geography and the spatialities of prison visiting: Visitation, recidivism and hyperincarceration. *Environment and Planning D: Society and Space* 31(1): 174–190.

Moran D (2015) *Carceral Geography: Spaces and Practices of Incarceration*. Farnham: Ashgate.

Moran D, Piacentini L and Pallot J (2012) Disciplined mobility and carceral geography: Prisoner transport in Russia. *Transactions of the Institute of British Geographers* 37(3): 446–460.

Muncie J (2009) *Youth and Crime*. London: Sage.

NSWLRC (2012) *Bail*, 13 June. Sydney: New South Wales Law Reform Commission.

Philo C (2014) One must eliminate the effects of … diffuse circulation [and] their unstable and dangerous coagulation: Foucault and beyond the stopping of mobilities. *Mobilities* 9(4): 493–511.

Sanders J and Grainger R (2003) *Youth Justice: Your Guide to Cops and Court in NSW*. Sydney: The Federation Press.

Shopfront Youth Legal Centre (2011) *Move-on Powers: Legal Information for Youth Workers*. Available at: www.theshopfront.org/documents/Police_move-on_directions.pdf

Shopfront Youth Legal Centre (2015a) *Bail: Legal Information for Youth Workers*. Available at: www.theshopfront.org/documents/Bail_2015.pdf

Shopfront Youth Legal Centre (2015b) *Police Powers and Your Rights: Legal Information for Youth Workers*. Available at: www.theshopfront.org/documents/Police_Powers_and_Your_Rights_May_2015.pdf

Stubbs J (2010) Re-examining bail and remand for young people in NSW. *Australian & New Zealand Journal of Criminology* 43(3): 485–505.

UK Children's Commissioner (no date) *Buzz Off Campaign*. Available at: www.childrenscommissioner.gov.uk/issue_rooms/buzz_off_campaign

Weatherburn D and Fitzgerald J (2014) *The Impact of the NSW Bail Act (2013) on Trends in Bail and Remand in NSW*, August 2015. Sydney: NSW Bureau of Crime Statistics and Research.

Weber L, Fishwick E and Marmo M (2014) *Crime, Justice and Human Rights*. London: Palgrave Macmillan.

White R (1990) *No Space of Their Own: Young People and Social Control in Australia*. Cambridge: Cambridge University Press.

Wong K, Bailey B and Kenny D (2010) *Bail Me Out: NSW Young People and Bail*. Sydney: Youth Justice Coalition.

Zedner L (2007) Pre-crime and post-crime criminology. *Theoretical Criminology* 11(2): 261–281.

5 Accommodation for asylum seekers and 'tolerance' in Romania

Governing foreigners by mobility?[1]

Bénédicte Michalon

Although the importance of mobilities has been emphasised in classic works regarding the prison, the social sciences have only recently explored them in respect of carceral procedures. The link between carcerality and mobility has been studied on several levels with regard to controlling foreigners. First, routes are increasingly structured by alternating free travel/forced displacement and are marked by somewhat long phases spent in detainment sites (Clochard, 2010; Michalon, 2012; Schmoll, 2014). Detention is also considered as an experience of stopping and waiting (Griffiths, 2014), or as an experience of route renegotiation (Darley, 2008). Research thirdly deals with the fact that confinement and circulation are structured through 'governmental mobilities' (Gill, 2009a), that is to say, displacements determined and organised by the institution, be it in connection with the ongoing procedure or with punitive aims (Michalon, 2013). They are an integral part of constraint and institutional control; they generate a sense of disorientation among displaced foreigners (Hiemstra, 2013), reducing the possibilities of external aid and upsetting internal sociability (Gill, 2009b). This displacement has theoretical implications. Closed settings become redistribution platforms for the flows of foreigners (Kobelinsky and Makaremi, 2008). They are therefore often conceived as part of a wider tendency: 'the global expansion of the confinement industry' (Martin and Mitchelson, 2009: 459) – confinement being understood as 'being kept in enclaves, whether they be concrete or abstract, ethnic, physical, legal or institutional' (Akoka and Clochard, 2015: 11). Lastly, mobility is no longer only an expression of freedom – a widely valued perspective in the *mobility turn* – but also an instrument of the power of the State, able to even force the issue (Baerenholdt, 2013; Creswell, 2006; Gill, 2013; Glick Schiller and Salazar, 2013; Turner, 2007).

Migration policies have increasingly resorted to the forceful placement of individuals in 'archipelagos' (Mountz, 2011) – including waiting at border zones, detention centres, police stations, and penal institutions. These policies also implement methods of control in so-called open environments. Drawing on a survey carried out in Romania, this chapter will examine two of these mechanisms designed to 'place' asylum seekers. The first consists of semi-open accommodation centres for asylum seekers and 'beneficiaries of protection' from the Romanian government. These centres are closed at night and open during the day. The second is called 'tolerance' (*tolerare* in Romanian) and consists of the temporary assignment of the right to remain on Romanian territory but not to obtain a residence permit. The accommodation of asylum seekers is presented as assistance provided by the State to persons in need of protection. Tolerance is intended for undocumented migrants who can neither be deported nor legalised. Although it is possible to consider that the ways in which these mechanisms limit the subjects' capacity to act, entry into these two schemes is done at the request of the foreigners concerned. They therefore raise the question of the nature and extent of the government's control over foreigners through management methods. First, there are those methods that occur prior to detaining a significant proportion of the rejected asylum seekers in detention centres; and second, those that usually occur after passing through detention – via tolerance. How and why does the Romanian government push foreigners to take part in their own management? The analysis of empirical data to follow highlights the central role of mobilities in the definition of governmentality of foreign nationals and in the way in which it is applied through these two mechanisms (Foucault, 2004). Contrary to detainment in detention, accommodation in asylum centres and tolerance are not intended to prevent mobilities other than those permitted by the government. Instead, they encourage foreigners to control and limit their travels themselves. These two mechanisms are therefore designed to guide and regulate behaviours – to control foreigners' room for manoeuvre with respect to certain aspects of their lives – by modulating the control exerted on them by two spatial vectors: residential localisation and mobilities.

The fall of Romania's leader Nicolae Ceauşescu resulted in a radical transformation of the configuration of migration and an expansion of immigration in 1990. External pressure was exerted on the Romanian authorities by international organisations (UNHCR and IOM) and by the States of the European Union for the country to overhaul its legislative and institutional system for managing migration. Tightened control over foreigners also reflects the political and administrative repercussions of the representation of Romania as a transit zone. This chapter draws on empirical data collected in 2009, 2010, and 2014. The General Inspectorate for Immigration (*Inspectoratul General pentru Imigrări*, IGI)[2] – the national institution in charge of immigration and asylum under the Romanian Ministry of the Interior – granted me authorisation to investigate three of the six centres for asylum seekers.[3] Data was collected through observations, informal discussions and semi-structured interviews. The survey conditions varied from one centre to another. I was sometimes able to visit the premises but without authorisation to speak with the people staying there, whom I contacted through associations providing assistance to foreigners. Elsewhere, the management authorised me to ask asylum seekers to voluntarily meet with me. I was sometimes able to move about freely by myself in the accommodation facilities. However, most of the research was conducted outside the centres, through volunteer work in a non-governmental organisation that specialises in helping foreigners, allowing me to conduct interviews with persons receiving aid from NGOs. I completed this work by interviewing the actors involved in the asylum management mechanism (senior managers or officers based in the centres) or those contesting it (actors in the voluntary sector, lawyers specialising in immigration law). In total, I was able to interview some 60 foreigners, 24 IGI officers, and 15 people working in NGOs and international organisations.

The governmentality of foreigners: The production of in-between subjects

The migration issue, including asylum, is now closely associated with the control of the borders and identified as a 'problem' in Romania. There is now an increasing focus on the fight against so-called illegal immigration (Michalon, 2015). Accommodation centres for asylum seekers and tolerance reflect this evolution. By tracing their major founding and operating principles, we are given a broad overview of what it means to govern foreigners without detaining them; we are also able to highlight the reasons for categorisation conveyed by these two mechanisms. Several decades ago already, Foucault demonstrated how 'dividing practices' can be used to distinguish subjects from each other and to allocate them to categories to then make them the subject of a specific institutional treatment (Foucault, 1994). The specific categories of subjects produced here are in-between categories.

Accommodating in centres to produce asylum seekers

Asylum seekers may have access to accommodation in one of six regional centres. Opened between 1995 and 2011, these centres can accommodate 1,150 people. The buildings belong to the IGI. Located in the capital and near the borders, they were set up in already existing buildings that were rehabilitated using national and/or European funding. The opening of these centres in Romania fits a broader trend at the European level, recognised since the early 1990s in different countries: that placing asylum seekers in ad hoc centres is a way of tightening asylum policies (Breen, 2008; Ghorashi, 2005; Kobelinsky, 2010; Szczepanikova, 2012; Stewart, 2005). Specialised accommodation makes it possible for authorities to exercise control over the applicants.

Contrary to what has prevailed elsewhere, in Romania, accommodation in centres is not mandatory while the asylum application is being examined; however, the Law on Asylum in Romania (hereafter LAR) outlines that applicants who do not have the necessary resources can be helped, through accommodation in a centre, the allocation of social housing or even through financial assistance for rent. The applicant must therefore request accommodation in the centres. This is authorised throughout the entire procedure. It can be extended a few months if the applicant has obtained a form of protection from the Romanian government – asylum under the Geneva Convention or subsidiary protection. People considered 'vulnerable' are favoured (elderly or disabled persons, unaccompanied minors, single parents and victims of torture). Access to the centres is therefore the combined result of a voluntary step taken by foreigners and the categorisation, prioritisation, and sorting process for foreigners by the administration.

State control is furthermore exercised by the multifunctional nature of these establishments, which are not only devoted to accommodation. They concentrate the range of tasks that the administration must provide to asylum seekers. They are sites that provide assistance, but they are also and above all sites for the administrative procedure of asylum. Applications are stored and processed there, regardless of whether the applicants are housed in the centre or outside of it. The geographical concentration of these administrative and social roles of the IGI is reflected in how the work of the agents who are assigned there is organised. 'Decision officers' (*ofiţer de decizie*) – police officers from the IGI who decide on asylum applications – perform the duties of decision officer, legal adviser for asylum seekers, and representative of the administration for appeals. This plurality of contradictory duties raises questions about the effectiveness of the assistance given to asylum seekers in case of conflicts of interest with the administration. It also gives decision officers an influential position in the centres. This organisation is prevalent everywhere except in Bucharest, where the accommodation centre is separated from the administration of the IGI. Daily interactions between applicants and decision officers are therefore much more limited than elsewhere, which may have consequences on the ways of imposing power and the fate of the migrants. Although assistance and supervision functions are intrinsically linked, as Robert Castel shows (1995), the organisation of the centres and institutional treatment of asylum seekers both serve to produce a population that will be categorised as 'vulnerable' and to place it under daily police control at the same time, while leaving the responsibility to be put in such a situation to the individual.

Generating a category of foreigners who can neither be deported nor legalised: 'Tolerance'

While accommodation in centres is offered to individuals on whom the administration must make a decision, tolerance itself refers to certain categories of undocumented migrants. Initially established in West Germany in 1965, this provision was implemented in Romanian law through close cooperation between the Romanian and German authorities in the 1990s.

> The measure of tolerance to remain in Romania ... means permission to stay in the country granted by the General Inspectorate for Immigration to foreigners who have no right of residence and who, for objective reasons, are not leaving the Romanian territory. For objective reasons ... refers to conditions beyond the control of the foreign national, which are unpredictable and cannot be prevented. (Emergency Ordinance No. 194 from 12 December 2002 on the Status of Aliens in Romania, hereafter EOSAR, art.102)

The expression 'objective reasons' covers several situations, such as having been released from detention, to have likely been a victim of human trafficking, or even to be subject to deportation but 'that sufficient evidence indicates it is not necessary to detain them in a detention centre' in order to remove them from the territory (EOSAR, art.103).

The attribution of tolerance involves the examination of certain conditions: the time spent in Romania, the provision of accommodation, family relationships in Romania, and the means of subsistence. In practice, tolerance is closely related to detention. According to police officials, this seems to be attributed to three main scenarios: if the detained foreigner is recognised as 'vulnerable', the attribution of tolerated status allows for his/her release from detention but does not annul the order of deportation; if the detained foreigner ultimately cannot be deported, he/she may be granted tolerance at the end of his/her period in detention; last, asylum seekers transferred to Romania as part of the Dublin agreements can avoid detainment if they accept the principle of 'voluntary' return to their country of origin.

Tolerance lasts for a period of less than six months and can be renewed until the reasons why it was attributed have disappeared. It is granted to people who will not leave Romania for a certain timeframe but does not necessarily represent a right of residence. Consequently, 'tolerated' migrants (*tolerați*)[4] will benefit from almost none of the rights afforded to migrants with documents – which explains what a tolerated Iraqi said in Bucharest in 2009, 'a tolerated person can stay for a short time to resolve his/her problem but he/she is not entitled to anything else except oxygen'. Tolerated migrants are obligated to frequently report to the authorities, the frequency is defined by the police officers themselves within a maximum of 60 days. These summons are often repeated a few days apart, reflecting the highly unstable and precarious nature of tolerance; 'it is only a period during which he/she [the foreigner] can find ways to leave the territory', admits an IGI officer in 2010.

Despite the fact that tolerance weakens rights, it is often the only option that the administration has left to manage certain individual situations. A quote from a police official in 2014 acknowledges this ambiguity:

> [I]n reality it [tolerance] is not used often because the risk of absconding is too great. It is granted to people who have been in Romania for many years, who were in an irregular situation but whom you could be sure would not leave [prior to their deportation]. This is for example the case of Syrians because they have no place to which they can return … This was also the case for an Iraqi man who was in Romania for seventeen years, it was clear that he would not run away. It is certain that he will return to Romania after his deportation, with a visa.

It is fair to ask if the existence of such a legal mechanism, which generates a category of non-deportable migrants, is intended only to subject migrants to tightened police handling.

While based partly on the free will of the subjects, asylum centres and tolerance create divisions among the foreigners and place some of them in specific categories of public action. Partly legitimised by the need to provide assistance to people that the administration have constructed as 'vulnerable', these mechanisms are also based on a special approach to the risk. 'Risk of absconding' – the alleged willingness to leave Romanian territory – is indeed a common pattern in these management methods, evoked repeatedly in the interviews with policemen. Considered a priori as presenting a 'risk of absconding', asylum seekers and undocumented migrants are subsequently the subject of police handling, the assumed goal of which is to prevent this 'risk'. This comes first by imposing control over the residence of foreigners.

Regulating behaviours through housing: Forced immobility

Asylum seekers and tolerated migrants settle in the housing of their choice, without losing the rights to which they are entitled. How, in these conditions, is housing a form of governmentality over foreigners? It appears to be an effective way for the State to regulate the behaviours of the subjects because it ensures control of (residential) localisation through bureaucratic practices. The manner in which the administration takes control of housing so as to control foreigners' room for manoeuvre varies depending on the population category concerned. 'The importance of socio-legal status in defining the housing-related rights and opportunities' (Dwyer and Brown, 2008: 204) is shown here. Legal status matters in the definition of control of foreigners through housing, since housing is supposed to guarantee and provide evidence that the asylum seekers and tolerated migrants are immobile.

Ensuring immobility: Bureaucratic control of residential localisation

Asylum seekers and tolerated migrants are monitored by the law enforcement authority by localising them through an accommodation address. This requirement is set by law for asylum seekers, regardless if they are housed in a centre or not. This restriction is justified by the needs of the procedure: court orders are either communicated directly in the offices of the IGI agents, or are sent by post to the last address given by the applicant. The law even provides for cases in which an asylum seeker housed in a centre would have been 'evacuated' and would have not given the new address to the authorities. For tolerated migrants, the law is vague and merely specifies that they must report any changes of residence. In practice, however, State officials require that foreigners declare their residence as soon as they request tolerance and that they bring proof of this housing via a lease to each meeting. This requirement is randomly modulated depending on the agents. While the administration generally accepts contracts for renting a bed-space, such as those given by an assistance association that manages a hostel, sometimes agents require a contract for an entire housing unit. The ambiguities of the law on this matter allow room, therefore, for discretionary practices. The administration tries to ensure foreigners' to limit their mobility through an immobile residence: we are here at the crossroads of the two major trends analysed by Jørgen Ole Bærenholdt: governmentality by mobility on the one hand, 'governmobility' – 'the self-government of people, governed through mobility, in more dynamic and relational ways involving material dispositions' (2013: 28) on the other.

Micropolitics of imposing immobility through the accommodation of asylum seekers

The governmentality of foreigners is also exercised within the accommodation itself. The micropolitics of imposing order are prevalent within centres for asylum seekers, which in particular include the control and regulation of circulations between the inside and outside and aim to reduce these movements. These establishments are semi-open: inhabitants can come and go during the day; however, circulation is impossible at night. No one can enter or leave without being checked and residents must leave their identification document – the temporary residence permit allocated to asylum seekers – at the checkpoint at each of the centre's entrances. Outside visits require approval from the watch officer and take place in visit rooms; visitors are not allowed to circulate throughout the entire building. The administration thus easily ensures the control of movement. At any time, its agents can see who is present or absent.

This first technique of control is completed in conjunction with the 'rounds' by the night duty officer, who counts all the residents in their rooms in the late evening. This means that occupants must leave the door open to be counted, something that some residents complain about, women in particular. Night rounds have a key role in supervision by the State: the internal regulations stipulate that if someone is missing when the police officer passes by, the foreigner is declared absent. After three nights of absence, the asylum seeker is 'evacuated'. Accommodation centres are used as a way to impose forms of immobility on asylum seekers: only those people who are actually on site can benefit from the assistance mechanism through accommodation.

Humanitarian accommodation or detention centre? Keeping tolerated migrants on site

The Romanian government does not have any specific accommodation for tolerated migrants. However, it has the possibility to use accommodation as a way to guide their behaviours: it negotiates with an association managing a hostel so that it is responsible for the accommodation of some of the tolerated migrants. Planned for about 20 people, this accommodation, organised through renting a bed-space, is renewed every month until the foreigner succeeds in obtaining a residence permit – or until they leave Romania. This renewal can occur for a maximum duration of roughly one year. The presence of the tenants is monitored daily by an employee of the association; missing persons are written off after a few days, and the usual interpretation of absences is that these people have crossed the border. The association also conducts legal and social assistance activities. It is precisely through these diverse competencies that the administration tries to put pressure on the association. In fact, for many people, being granted a tolerated status depends on assistance from the humanitarian sector. The status of 'vulnerable' is mainly recognised by the authorities when they have to manage the fate of people who are the focus of militant activism. Yet, accommodation is part of the terms of negotiation:

authorities often require that the association provides accommodation for certain tolerated migrants, families in particular, and they threaten to not allocate tolerance and to detain the persons concerned if they are not provided accommodation. Thus the administration seeks to ensure a form of control over tolerated migrants through the association, the humanitarian priorities of which are put in a difficult situation or even take second place to the administration's priorities. Here we see how the assumption of 'risk of absconding' influences the institutional treatment of tolerated migrants: they must remain under control, either through their accommodation in hostels or their detainment. Mechanisms that do not guarantee immobility, or monitor those persons, are rejected by the authorities.

If the governmental control of foreigners requires producing categories of specific subjects in order to be implemented, it also requires, as Jonathan Darling writes, 'to accommodate those categorised as suspicious subjects' (Darling, 2011: 267). The institutional treatment of asylum seekers and tolerated migrants in Romania shows how accommodation becomes a method of immobilisation. This can be explained by the fact that voluntary mobility, primarily seen as a 'risk', discredits foreigners and helps to weaken or even question his/her legal status. This hypothesis is supported by the analysis of techniques used to control the movements of foreigners.

Controlling foreigners through a *continuum* of mobilities

Mobility is evident through the request for asylum and tolerance: sometimes mobility is free but geographically limited by the law, and it is sometimes possible, if only in a binding manner, by the authorities. Institutional supervision in open settings therefore does not maintain a continuous opposition between carceral mobilities and 'free' mobilities, but rather a *continuum* of practices to govern foreigners through circulation, which ranges from forms that are not well regulated by the administration but which involve the foreigner accepting and participating in his/her own control, to governmental mobilities geared towards meeting the administration's goals.

Mobility limited

The control of asylum seekers and tolerated migrants involves a restriction of their freedom of movement. For asylum seekers, mobility is restricted to the perimeter around the place of residence. More distant travel is subject to approval by the IGI, for which notice will be given after 'an individual, objective and impartial analysis' of the request (LAR, art.19). Authorisation to travel is granted for a specific destination and for a fixed duration. The goal of this technique, common in the governmentality of all asylum seekers, is to make it easier for the administration to locate them; but its goal is to also demonstrate the power of the State over the subjects and to make the presence of this power felt.

A greater restriction exists for the tolerance regime: in reality, it is only valid within the jurisdiction of the territorial formation of the IGI that has attributed the status of tolerance.[5] Foreigners recognised as tolerated migrants in Bucharest are not tolerated elsewhere in the country – which immediately limits the benefits of any mobility in Romanian territory. Thus, tolerated migrants are once again considered illegal if they cross the borders of the department in which they obtained their tolerated status. This restriction also means that any travel outside the attributed tolerance zone is subject to prior authorisation by the IGI. Although not detained, they are confined. The geographic restriction of tolerance and the limited right to mobility shows a *continuum* of government mechanisms via the marginalisation of foreigners.

The effective implementation of these measures rests first and foremost on consent from foreigners. The public force then focuses its attention on the foreigners' intentions, as an IGI agent said in 2010: 'it is necessary to pay attention to those who try to pass the border ... For the moment we are still a country of transit. But it is more obvious in the centres close to the [Western] border'. Any request for travel is considered a priori as being part of a plan to leave Romania and State officials scrutinize certain behaviours considered as indicators of 'risk of absconding'. Repeated absence from the centres is one such behaviour. Possible arrest outside of the area of residence is another, as an official of the IGI explains in 2014: 'If we find them in Constanța, we turn a blind eye, however if we find them in Timișoara this means that they want to leave so this is not good at all. We are not idiots'. The distinction introduced here links back to Foucault's claims of circulation as a security objective (Foucault, 2004).

Governmental mobilities to isolate asylum seekers

Detainment needs mobility to be implemented and lived by foreigners, as evidenced by the governmental mobilities imposed on foreigners detained in detention centres (Michalon, 2012). Similar mechanisms exist for the government of asylum seekers. Management of asylum seekers in centres is partly based on mobilities that are decided and organised by the administration and that includes, as in detention (Michalon, 2013), both the procedure as well as forms of punishment. Although applicants who are already settled in Romania can request a place in the centre nearest to their place of residence, newcomers are distributed according to the administrative needs, and solely on this basis. In interviews, the police hierarchy first and foremost emphasises the primacy of the territorial criterion – there is a distribution map for asylum seekers in the accommodation centres. However, other considerations seem to prevail; foreigners are distributed according to practical criteria – the availability of places in the centres, but also according to political criteria: they are grouped by nationality or groups of alleged ethnic, cultural or religious proximity. Police work thus puts great emphasis on categorisation mechanisms that define asylum seekers by their alterity. Moreover, distribution occurs between the time the asylum application is lodged and the interview with a decision officer. Foreigners are not informed in advance of their possible transfer to a centre, which is organised by State agencies.

The goal of these displacements may also be to segregate people considered as dangerous: the law specifies that 'for reasons justified by public interest, national security, order, health and public morality, the protection of the rights and freedoms of others – even if the foreigner has the material resources required for his/her care – the IGI can … proceed with his/her displacement' (LAR, art. 17). Asylum seekers acting as intermediaries are among those targeted by this measure. The way to govern asylum seekers involves controlling their mobility which at the very least is intended to cause social isolation, which introduces an obvious similarity with the treatment of prisoners (Svensson and Svensson, 2006).

The objective of institutional monitoring in an open setting is not to stop mobilities – whereas various authors see it as the first rationale for penal institutions such as prisons or the psychiatric asylum (Martin and Mitchelson, 2009; Philo, 2014). Nevertheless, circulations and their control play a crucial role in open institutions, with variations in how coercion is exercised, either by direct action by State agents, or by acceptance from foreigners and their participation in their own control through behaviours that are in line with what the State expects of them. If this is not the case, punitive measures are taken that will affect the mobility of the foreigner.

Punitive mobility to guarantee the governmentality of foreigners

The State's control over asylum seekers and tolerated migrants becomes punitive when the law is infringed or the centres' regulations are broken. Compliance with the constraints imposed on mobility or immobility is central here: it is foremost on this level that the behaviours of foreigners are evaluated by police forces. The law is somewhat precise on the punitive measures to be taken and gives IGI officials a wide margin of discretion.

Punishment by exclusion from the assistance mechanism for asylum seekers

Asylum seekers accommodated in a centre must comply with internal regulations. Although the proposed sanctions are left to the discretion of the personnel in the establishments, playing on 'closing-opening' rules may be subject to punitive actions. Residents are expected to be back before closing time; if not, they are forced to spend the night outside the centre. Younes, who stayed in an establishment in the Eastern part of the country for several months, said in 2010 that life in the centre was punctuated by altercations generated in part by these operating rules. Absences may also be subject to punitive interventions by State agents. National law authorises managements to 'evacuate' residents who are absent too often. An 'evacuation' report is made and noted in the applicant's administrative file. As a warden explained in 2010:

> [T]he evacuation is done only on paper; we put it on record because it may be that the applicant is trying to leave the country but cannot do so and therefore

comes back here. He/She may then be relocated for the remainder of the procedure.

Although some agents give foreigners ample freedom and tolerate lateness and/or absences, others go beyond what the law requires. One warden reported that if an asylum seeker is absent for more than three days, he informs the IGI and the Romanian Border Police (*Poliția de frontieră*) that someone has 'escaped' and that he/she might try to cross the border. Being left to decide which measures to take to cope with violations of the establishments' regulations opens the door to discretionary practices by the agents.

A second level of punitive measures deals with non-compliance of the territorial limitation of mobilities of asylum seekers. Arrest outside of the area of residence or even crossing State borders will be, similar to an 'evacuation', noted in the asylum seeker's file. Violation of the order to remain immobile may affect the current procedure. Any violation of the travel restrictions will be considered as proof that the applicant tried to leave Romania while his/her asylum application was being reviewed, an element that will be used against him/her because it supposedly implies the 'non-sincerity' of his/her request – even if the administration has no information confirming this hypothesis. The assumption of 'risk of absconding' is integrated in the management of centres and helps the administration not only to distinguish the 'good' asylum seekers from the 'bad', but also to separate the 'true' asylum seekers, who 'merit' protection from the Romanian government, from the 'fake'.

Punishment by confinement of tolerated migrants

Punitive measures are set by law and are stricter for tolerated migrants. In particular, two types of offences are targeted: failure to comply with the obligation to regularly report to the authorities and leaving the allocated zone of tolerance. Since these two types of violation of the law are interpreted as evidence of willingness to illegally leave Romania, foreigners are sanctioned by placement in a detention centre. Rustam, whom I met in 2010, was granted a tolerated status after spending six months in detention. When he arrived two years earlier in Romania, he filed an asylum application, obtained a place in a centre and then was transferred to another. Several weeks later he left Romania and filed a new application for asylum in Belgium. Returned to Romania by the Dublin agreements, he was once again placed in detention. As he could not be deported, he was given tolerated status when he left the detention centre. A few weeks later he was arrested while trying to leave the country. As a result he lost his tolerated status and was held in detention for not complying with the obligations to present himself to the IGI and for not reporting any travel plans. The founding ambiguities of tolerance are revealed here: initially intended for persons who cannot leave Romania, tolerance is based on the immobilisation of tolerated migrants and punishes those who travel.

Noted in individual files, punitive measures have an impact on the legal status of foreigners. Although asylum applications and tolerance are, in some respects, less stringent than detainment, the non-alignment of the behaviours of foreigners to what is expected of them by the administration questions their very presence in Romania. The control of migrations involves the standardisation of behaviours and punishment.

Conclusion

This chapter has analysed the role of mobility in the governmentality of foreigners through two control mechanisms prevailing in Romania: accommodation centres for asylum seekers and the mechanism of 'tolerance'. Contrary to governance by detention, the control over asylum seekers and the tolerated foreigners is grounded not on a principle of forced restriction of mobility, but on a principle of incentive to the self-limiting of movement by the migrants. Categorised by the authorities as 'vulnerable' and regarded as presenting a 'risk of absconding', asylum seekers lodged in centres and tolerated migrants are consequently the object of a police action, who demand from foreigners that they limit their own mobility or even that they stand still. However, constraints to mobility appear in the form of punishment if the formal and informal rules placed upon foreigners are infringed. Such punishment has consequences on the legal status of the foreigners and can lead to their detention.

What can we learn though, from the governance of foreigners in this context? Control through asylum seekers' centres and the mechanism of tolerance raises the issue of the individualisation and the contractualisation of migration policies, since the effectiveness of the analysed mechanisms depends on how foreigners accept them and how their behaviours conform to the norms prescribed by the law and the police officers. The so-called open means of controlling migrants consequently seem to share common points with certain types of probationary penal measures and could develop in the coming years, since States are urged to develop so-called alternatives to detention, such as house arrest or electronic tagging. Consequently, the analysis of the regimes of governmentality of foreigners – through mobility measures – raises questions about the classical distinction between open and closed facilities. In the penal field, this long standing debate is translated by an opposition between, on the one hand, approaches that assume a difference in nature between the probationary sanctions and the penal sanction; and, on the other hand, approaches that plead for a reading in terms of continuity between open and closed settings leading to the idea of the extension of the penal net, developed by Michel Foucault and David Garland in particular (De Larminat, 2014). The situation observed in Romania allows us to conclude that open and closed institutions work together in the governmentality of migrants, the various types of mechanisms intersecting in migration control and in the paths of migrants that follow.

Notes

1 In memory of Matthieu Giroud, longtime friend, shot in the Bataclan in Paris on 13 November, 2015.
2 This institution succeeded in 2012 the Romanian Office for Immigration (*Oficiul Român pentru Imigrări*, ORI), founded in 2007. By facility of reading, I will refer to the current name (IGI).

3 In order to keep the anonymity of my interlocutors, the places where the interviews were conducted and the functions of interviewees are not specified; all the names were changed.
4 Although, for convenience, I am using the official terms of 'asylum seekers' and 'tolerated' migrants, I remain aware of the homogenising and essentialising effects of these categories.
5 The IGI is divided into 41 'territorial formations' which correspond to the 41 departments in the country.

References

Akoka K and Clochard O (2015) Régime de confinement et gestion des migrations sur l'île de Chypre. *L'Espace politique* 25: 6–15. Available at: http://espacepolitique.revues.org.gate3.inist.fr/3381
Bærenholdt JO (2013) Governmobility: the powers of mobility. *Mobilities* 8(1): 20–34.
Breen C (2008) The policy of direct provision in Ireland: A violation of asylum seekers' right to an adequate standard of housing. *International Journal of Refugee Law* 20(4): 611–636.
Castel R (1995) *Les métamorphoses de la question sociale: Une chronique du salariat*. Paris: Fayard.
Clochard O (2010) Le contrôle des flux migratoires aux frontières de l'Union européenne s'oriente vers une disposition de plus en plus réticulaire. *Carnets de géographes* 1: 6–15. Available at: www.carnetsdegeographes.org/carnets_recherches/rech_01_03_Clochard.php
Creswell T (2006) *On the Move: Mobility in the Modern Western World*. Abingdon: Routledge.
Darley M (2008) *Frontière, asile et détention des étrangers. Le contrôle étatique de l'immigration et son contournement en Autriche et en République tchèque*. Unpublished doctoral dissertation, Institut d'Etudes Politiques de Paris.
Darling J (2011) Domopolitics, govermentality and the regulation of asylum accommodation. *Political Geography* 30(5): 263–271.
De Larminat X (2014) *Hors des murs. L'exécution des peines en milieu ouvert*. Paris: Presses universitaires de France.
Dwyer P and Brown D (2008) Accomodating 'others'? : Housing dispersed, forced migrants in the UK. *Journal of Social Welfare & Family Law* 30(3): 203–218.
Foucault M (1994) Le sujet et le pouvoir. In: Foucault M *Dits et écrits II. 1976–1988*. Paris: Gallimard: 1041–1062.
Foucault M (2004) *Sécurité, Territoire, Population. Cours au Collège de France (1977–1978)*. Paris: Gallimard.
Gill N (2009a) Governmental mobility: The power effects of the movement of detained asylum seekers around Britain's detention estate. *Political Geography* 28(3): 186–196.
Gill N (2009b) Longing for stillness: The forced movement of asylum seekers. *MC Journal: A Journal of Media and Culture* 12(1): 10–15. Available at: www.journal.media-culture.org.au/index.php/mcjournal/issue/view/still
Gill N (2013) Mobility versus liberty? The punitive uses of movement within and outside carceral environments. In: Moran D, Gill N and Conlon D (eds) *Carceral Spaces: Mobility and Agency in Imprisonment and Migrant Detention*. Farnham: Ashgate: 19–35.
Ghorashi H (2005) Agents of change or passive victims: The impact of welfare states (the case of the Netherlands) on refugees. *Journal of Refugee Studies* 18(2): 181–198.
Glick Schiller N and Salazar NB (2013) Regimes of mobility across the globe. *Journal of Ethnic and Migration Studies* 39(2): 183–200.
Griffiths M (2014) Out of time: The temporal uncertainties of refused asylum seekers and immigration detainees. *Journal of Ethnic and Migration Studies* 40(12): 1991–2009.

Hiemstra N (2013) 'You don't even know where you are': Chaotic geographies of US migrant detention and deportation. In: Moran D, Gill N and Conlon D (eds) *Carceral Spaces: Mobility and Agency in Imprisonment and Migrant Detention*. Farnham: Ashgate: 57–75.

Kobelinsky C and Makaremi C (2008) Éditorial. Le Confinement des étrangers: Entre circulation et enfermement. *Cultures et Conflits* 3(71): 7–29.

Kobelinsky C (2010) *L'accueil des demandeurs d'asile. Une ethnographie de l'attente.* Paris: Editions du Cygne.

Martin LL and Mitchelson ML (2009) Geographies of detention and imprisonment: Interrogating spatial practices of confinement, discipline, law, and state power. *Geography Compass* 3(1): 459–477.

Michalon B (2012) La mobilité au service de l'enfermement? Les centres de rétention pour étrangers en Roumanie. *Géographie et Cultures* 81: 91–110.

Michalon B (2013) Mobility and power in detention. The management of internal movement and governmental mobility in Romania. In: Moran D, Gill N and Conlon D (eds) *Carceral Spaces: Mobility and Agency in Imprisonment and Migrant Detention*. Farnham: Ashgate: 37–55.

Michalon B (2015) Invention du 'problème' de l'immigration et reconfigurations frontalières en Roumanie. In: Vayssière B (ed) *Penser les frontières européennes au XXIe siècle. Réflexion croisée des sciences sociales*. Bruxelles: Peter Lang : 217–230.

Mountz A (2011) The enforcement archipelago: Detention, haunting, and asylum on islands. *Political Geography* 30(3): 118–128.

Philo C (2014) 'One must eliminate the effects of … diffuse circulation [and] their unstable and dangerous coagulation': Foucault and beyond the stopping of mobilities. *Mobilities* 9(4): 493–511.

Schmoll C (2014) Gendered spatialities of power in 'borderland' Europe: An approach through mobile and immobilised bodies. *International Journal of Migration and Border Studies* 1(2): 173–189.

Svensson B and Svensson K (2006) *Inmates in motion: Metamorphosis as governmentality: a case of social logistics*, Lunds, Lunds University, Socialhögskolan, Working paper series (5): 10–15. Available at: http://lup.lub.lu.se/luur/download?func=downloadFile&recordOId=533245&fileOId=625397

Szczepanikova A (2012) Between control and assistance: The problem of European accommodation centres for asylum seekers. *International Migration* 51(4): 130–143.

Stewart E (2005) Exploring the vulnerability of asylum seekers in the UK. *Population, Space and Place* 11(6): 499–512.

Turner B (2007) The enclave society: Towards a sociology of immobility. *European Journal of Social Theory* 10(2): 287–303.

Laws

Emergency Ordinance No. 194 from 12 December 2002 on the Status of Aliens in Romania, *Official Journal of Romania n°421 from 05 July 2008* [Ordonanţa de urgenţă a Guvernului nr.194 din 12 decembrie 2002 privind regimul străinilor în România, *Monitorul Oficial numărul 421 din 05 iulie 2008*] (EOSAR).

Law No. 122 from 4 May 2006 on Asylum in Romania, *Official Journal of Romania*, No. 428 from 18 May 2006 [*Legea nr.122/2006 privind azilul în România, Monitorul Oficial numărul 428 din 18 mai 2006*] (LAR).

Part II

Circulation

v. movement in a circle, circular motion or course

6 'Doing time' differently

Imaginative mobilities to/from inmates' inner/outer spaces

James Gacek

Introduction

A major contribution of carceral geography has been its suggestion that the 'carceral' is something more than merely the spaces in which people are enclosed. Instead, this body of work insists that 'carceral' is a psychological and social construction of which the effects of imprisonment can leave lasting impacts upon inmate bodies (Moran, 2015). Certainly, the embodiment of imprisonment moves beyond the physical space of the prison. While carceral geography can make a distinctive contribution to examining the prison's impact on inmates' lives both while 'doing time' and post-release, this chapter investigates how inmates attempt to lessen the impact of carceral regimes by reverting inwards to escape the prison. In other words, understanding how and why inmates move from the real to the imaginary can illuminate discussions surrounding the conditions of prison life they routinely endure.

The notion of the 'carceral' as a social constructive process existing both separate from and within spaces of incarceration is central to this chapter. The assumed binary between the 'inside' and 'outside' of prisons has been critiqued extensively (for example, Baer and Ravneberg, 2008; Farrington, 1992; Moran, 2015). Generally, such arguments have posited that conceiving the binary distinction of 'inside' and 'outside' suggests that prisons are spaces 'outside of and different from other spaces, but still inside the general social order' (Baer and Ravneberg, 2008: 214). Indeed, it is not a far cry to depict such prison spaces as 'real places, actual places, places that are designed into the very institution of society' (Foucault, 1998: 178). Prisons, as 'not-so-total' institutions (Farrington, 1992), can exist as enclosed spaces with an identifiable-yet-permeable membrane of structures, policies and mechanisms, all of which typically maintain 'a selective and imperfect degree of separation' between what lies beyond and what exists inside prison walls (Moran, 2015: 90). However, whilst carceral geography has analysed the mobility of inmates moving from society to prison and back again, this chapter serves to fill the research lacunae that exists between the inmate's movement between the psychological and social levels of being. In sum, the mobility of inmates to psychologically enter the inner spaces of their minds to avoid and distance themselves from the prison life that exists 'outside' their anatomical control requires a greater focus.

In this chapter, I explore the complex comprehensions and meanings behind inmate mobilities in terms of one's movement between the real and the imaginary within carceral spaces, what imagination can provide for inmates, and how incorporating inmates' minds and 'mental spaces' into the analysis enriches spatial and mobilities literature. This analysis is supported by semi-structured interviews conducted with ten former inmates living within the city of Winnipeg, Canada, identified and recruited using snowball sampling. Two men interviewed identified as Caucasian, two men identified as Asian, and six men identified as Indigenous.[1] All of the men had experienced incarceration within correctional institutions in Manitoba.[2]

The John Howard Society of Manitoba, a social service agency located in Winnipeg's inner city, provided a small office where these meetings between the respondents and myself could take place. Access to these men for both recruitment and interviewing proved difficult. Within the sample of ten men I acquired, I had to rely on two of the ten men to stand in as my representation for recruitment when they were speaking with other men that were interested in my project. It was through these men's contacts and *connexions* that I was able to achieve my sample. Additionally, the purpose of the study was to generate 'context-dependent knowledge' pertaining to the impact of incarceration on identity construction through such narratives (Flyvbjerg, 2006: 222). This chapter comprises some of these findings.

To offer a more nuanced understanding of the intricate relationship between space, mobility and imagination, I first consider the theoretical significance of Simmel's and Urry's respective works for spatial and mobility thinkers seeking novel ways to study spaces of incarceration. Following this, I discuss how even the 'gloom' of carceral spaces can produce meaningful, distinct experiences for inmates that invigorates the imagination. Especially within the context of Canadian prisons, due to the overrepresentation of Indigenous people behind these walls, incorporating Indigenous spirituality and sacredness, in particular, enriches such discussions of imaginative transcendence. I then draw on responses from several of the men interviewed to show how activities that incorporate either play or rest in prison can enliven men's bodies and strengthen their power of conception within the imaginary of the mind. I then analyse how spirituality can build paths to phantasmagorical-albeit-ephemeral spaces. Sacred spaces in the prison not only assist Indigenous inmates in their healing and rehabilitation, but also forge a *connexion* to the divine that heightens spiritual transcendence. In its conclusion, this research suggests that mobility and imagination provides rich analytical possibilities for carceral geography, and offers a conceptual entry point for criminologists and geographers alike to study and interrogate key dilemmas of spaces of incarceration.

'Moving' differently: Simmel, Urry, and 'imaginative travel'

Simmel (1997) provides a unique interpretation of the significance of mobility, in which he notes the exceptional human achievement involved in creating a 'path' that links two particular places. While he argues that only in 'visibly impressing the path into the surface of the earth that the places [are] objectively connected' (Simmel, 1994: 6), there is still a necessity for people to travel backwards and forwards between the places and 'subjectively' connect them in their minds, regardless of how often this travel occurs. Such impressions of path-building through imagination create a permanent *connexion* (following Simmel's French terminology) between these places. In effect, the 'will to *connexion*' became one of the greatest (and specifically) human achievements, as the ability to summon imagination and creativity had shaped and built paths that connected the phenomenological experiences of the person existing on earth with the inner spaces of the mind. This achievement of *connexion* reaches its zenith with a bridge that symbolises 'the extension of our volitional sphere over space' (Simmel, 1994: 6). It is only through humans' will to *connexion* that the banks of a river are not just apart but separated, and therefore 'bridge-able' (Urry, 2007). Humans, then, are able to 'see' these *connexions* in their 'mind's eye' as separated *and* as thus needing *connexion*. Simmel summarises the power of 'conception', or human imagination: 'if we did not first connect [places] in our practical thoughts, in our needs and in our *fantasy*, then the concept of separation would have no meaning' (Simmel, 1994: 6, italics added).

Undoubtedly, Urry's focus on 'imaginative travel' is quite substantial in understanding the *connexion* between our practical thoughts and the fantasies that exist within our consciousness. He argues that there are many different forms of imaginative travel that enable people to 'travel' elsewhere through their texts, guidebooks and brochures, photos, postcards, and memories. In some cases, such travel can substitute for physical travel (Urry, 2007, 2008, 2010, 2012). However, imaginative travel seems to generate both the desire for travel and for being bodily within other places. Indeed, an emphasis on the imaginative aspect of mental psyches is pivotal to understanding the imaginative travel of inmates, as such 'travel' merely requires these men to sensually disengage from their physical carceral containments. As discussed below, imaginative episodes are sparked when inmates reengage with activities that transfer their consciousness to something other than the stressful and mundane reality they endure. The need to escape the mundane reality and transcend into more pleasurable experiences and imaginary spaces is ever present. However, situating this prevalence in terms of the overrepresentation of Indigenous peoples within Canadian correctional systems suggests that we then must explore how Indigenous identities are (re)created by imaginative mobilities and interrogate how Indigeneity is conceptualised within such spaces.

Space and Indigeneity

The 'spatial turn' has been 'one of the most recently proclaimed turns within the human and social sciences', and far from being a unified movement, the thematic and methodological reorientation consists of a variety of often diverse approaches to understand and conceive space (Frank, 2009: 66). Spatiality has become more prevalent within prison sociology and carceral studies (Crewe et al., 2014; Finanne and McGuire, 2001; Hogg, 2001; Jewkes, 2013; Milhaud and Moran, 2013; Moran, 2012, 2013a, 2013b; Moran et al., 2013). Indeed, the spatial arrangement of the prison has always been instrumental in maintaining control and organisational order, arguably implemented through 'the generation of an aesthetic, or feeling of, efficiency and the exclusion of disorder and distraction' (Hancock and Jewkes, 2011: 616; see also Kersten and Gilardi, 2003). Rhetorically, even Foucault (1977: 228) asks: 'is it surprising that prisons resemble factories, schools, barracks, hospitals, which all resemble prisons?' This sensual disengagement, as an internal aesthetic of the prison system, is itself oriented toward the organisation of behaviour. Sensual disengagement plays a significant role in the precognitive sensitisation to the orderly and the rational and a desensitisation to that which requires senses, such as activities, entertainment, and other modes of distraction. As aesthetic statements of state and sovereign power, prisons have become 'colossal enclosures of space, every bit as striking and cavernous as their burgeoning industrial counterparts' (Hancock and Jewkes, 2011: 616). While such an 'assault on the senses' could occur from the physical prison environment itself (Hancock and Jewkes, 2011: 627; see also Bowker, 1980; Tataro, 2006), new generations of prison design have increasingly come to encourage and facilitate communal interactions, which might be a welcome antidote to the 'disenchanting, estranging, and sometimes confrontational designs of traditional prisons' (Hancock and Jewkes, 2011: 627). Nevertheless, when examining the links between prison designs it appears that what has still survived throughout is the boredom and mundanity of carceral spaces, coupled with the sensual disengagement that is felt by inmates on a daily basis.

Such sensual (dis)engagement can be witnessed in the work of Tim Edensor, particularly in terms of how he reimagines the relationality of artificial lighting within gloomy and dark spaces. Indeed, 'the melding of illumination and darkness has a unique capacity to transform space and generate atmospheres' (Edensor, 2015a: 436; see also Edensor, 2013). According to Edensor, although experiences and rhythms 'play out differently according to geography, all sighted people perceive, sense, act and construe meanings of space, place, and landscape according to diverse, changing qualities of luminosity and murkiness' (2015b: 559). However, whilst the Western prison model still exudes varying degrees of gloom for its inhabitants, hope remains for inmates housed in such spaces. According to Edensor (2015a: 432), relative to the space, gloom has the power to '[enliven] the body, [sharpen] the senses and [make] one aware of others, producing a heightened, tactile sense of mobility'.

Indeed, the survival of gloomy and mundane carceral spaces has not been immune in the Canadian context. Comack (2008) has addressed the racialised character of the prison system, noting that prison in Canada has become the contemporary equivalent for many young Indigenous people of what the Indian residential school represented from the early years of Confederation into the twentieth century (Comack 2008: 12). Indeed, speaking in terms of the white settler colonialism that has, and still, shapes Canada's history, Canadian prisons have been but one space of confinement within a long history of Indigenous incarceration (Gacek, 2015). Aside from the burgeoning population of Indigenous peoples in prisons, other spatial forms of this incarceration include Indigenous reservations, residential schools, and the foster homes Indigenous children inhabit as they too are over-represented in the Canadian foster care system (Woolford and Gacek, 2016). Therefore, conducting research that incorporates Indigeneity should matter to spatial and mobility thinkers, as Indigenous experiences have the power to challenge dominant Western understandings of mobility and incarceration, while simultaneously opening up alternative knowledges of how this doubly marginalised group – as both Indigenous and incarcerated – apprehends their move from the real to the imaginary. Furthermore, such research must always be sensitive to the broader understandings and contentions surrounding many Indigenous cultural heritages – specifically, questioning how power, resistance, culture, and authorship interweave into spatiality at play. In order to configure an understanding of 'Otherness' within carceral spaces, we must remain aware of the intimate entanglement of 'distinct colonial histories and imperial institutions of power' (Abu El-Haj, 2013: 56), especially from a Canadian context of settler colonialism.

Moreover, McKegney contends that by investigating what it means to be 'Indian', he observes a means of tracking 'colonial simulations and technologies of coercion that have served to alienate Indigenous men from tribal specific roles and responsibilities' (McKegney, 2014: 3). Community reintegration, heteropatriarchy, assimilation into dominant White cultures, and Indigenous continuance are just a few struggles that Indigenous men-as-former-inmates continue to face. However, as the next section of this chapter demonstrates, inviting these men to speak will not simply voice their concerns to a broader audience; rather, their words in concert with this chapter contributes to the growing concern to heed the call to ameliorate Indigenous peoples' unfortunate circumstances (Comack 2008; Gacek, 2015).

'Imaginative' mobilities

'Nobody wants to be [in prison] … As fast as you get in there you want out of there' (Eric[3]).

A recurring theme throughout the study was a desire to move away from the carceral space. The men indicated several activities that allowed them to endure the boredom of prison. The two main activities were sports and gym workouts, as well as playing games such as chess, checkers, and card games. Other pursuits included watching television, sleeping and napping, reading, and writing letters to loved ones. As Issac suggested, activities such as these '[take] your mind [to] a different place', and ultimately, this travel to the mind space 'makes you think differently' and secures your sanity within the prison.

Gym access and sports were substantial to the well-being of these men (Gacek, 2015). Indeed, lifting weights and playing sports was crucial to alleviating the mental stress often felt while surviving prison life. The gym was beneficial for

Chris, as 'it takes a lot of stress out [of me], especially when [I was] playing sports ... like [American] football and hockey'. Kupers documented similar experiences with the inmates he interviewed, where he argued that sports and fitness activities in prison 'engage men's minds and bodies to varying degrees and, in the process, help them do their time' (2001: 62). Such activities become a crucial survival strategy, a practise that is intended to 'create and maintain ... mental health in a hostile, unhealthy place' (Kupers, 2001: 63). Like Chris's example, participating in sports and gym time helps inmates displace mental frustration and anger, allowing the rage out of their bodies and psyches before it explodes or turns in on them.

Furthermore, as Eric explained to me, playing games like chess, checkers, or canasta[4] was essential to 'burn hours and hours' of time away. For Eric, card games like canasta were enjoyable as 'it takes about an hour minimum to just play one game ... so, there would be like ... four, five, six games a day'. Similarly, watching television shows and films in prison distracted the men from the prison life they endured. Henry affirms this notion: '[It] did distract me from being [there] ... there's nothing else to do, you know, I'm bored, I'm just going to sit down and twiddle my thumbs or just try to go work out ... but I think movies really helped'. Additionally, it has been noted that the spatial and temporal disciplining of sleep is an important aspect of the power/knowledge nexus (Williams, 2005). However, my findings suggest an inherent paradox with sleep in the prison. While inmates must follow the routine of when and where they sleep and its duration, concomitantly these very activities render inmates the ability to travel through consciousness and, in the state of dreaming, achieve a sort of personal liberation within their imagination – a mobility and liberation that is hardly granted in their mundane prison realities. The monotony of the prison setting was indicated in Dave's experiences, for even if he slept to escape the chaos of the prison population, eventually '[you] just wake up, and you're still in the same place, [still] locked up' as before. Therefore, the gloomy prison space evokes and mobilises a desire for imaginative removal, all the while imposing the condition that returning to the dark reality will occur once inmates awaken from their slumber.

Simmel not only argued that 'space in general is only an activity of the mind', but also observed: 'It is not the form of spatial proximity or distance that creates the special phenomena of neighbourliness or foreignness ... Rather, these ... are facts caused purely by *psychological content*' (Simmel, 1997: 137–138, italics added). Indeed, the human body is transformed once 'proximity and connectivity are *imagined* in new ways' (Hannam et al., 2006: 2, italics added). Similarly, Urry indicates that the advent of 'diverse mobilities', such as imaginative travel, has recreated our understanding of 'inner mobility', for which 'coming and going, being both here and there at the same time' (2010: 348) has become more significant than ever before. Inner mobilities are not solely apprehended in terms of various globalised networks and flows, as Urry suggests; my findings indicate that a consideration of the abstract and surrealistic spaces that exist within human consciousness and cognition is warranted. Imaginative travel involves either the experience or the anticipation in one's imagination of the 'atmosphere of place' (Hannem et al., 2006: 14). The corporeal bodies of inmates, as affective vehicles, become attuned to place and movement within the prison. Subsequently, the mundane and gloomy atmospheres construct alternative emotional and imaginary geographies to navigate: 'When gloom descends … the varying levels of [gloom and murkiness] … motivate or restrict the body's movement, and shape the emotional affective response to space' (Edensor, 2013: 453). Therefore, if inmates, as people, are 'enmeshed in social dramas wherein actions depend upon negotiation, approval and feelings' (Larsen et al., 2006: 268), then there are grave social and emotional consequences when these men become connected to and embedded within complex inmate networks. Therefore, the use of imagination by inmates is crucial when adapting to carceral spaces, as instances of imagination provide inmates the capacity to emotionally adjust to incarceration, while simultaneously reviving and recalibrating the inner self.

Sacred spaces and spiritual transcendence

'I felt a feeling I never felt before … like I was being reborn' (Adrian).

Within society, people and places that are racialised or otherwise stigmatised are both literally and figuratively erased from the official landscape (Foote, 2003; Kobayashi and Peake, 2000; Price, 2010). However, one way in which to reclaim space for marginalised groups within prison – that is, the constructed, geographical site of incarceration and punishment – is to reimagine the sacredness of cultural heritage. Indeed, sacred space can take on a 'doubled quality' in which the space can be intimate, protective, sheltering, and inhabited, while concomitantly conditions the possibility 'of openness to the cosmos' (Game, 2001: 236). It is both centring and opening to religious and spiritual experiences. Sacred spaces mark out a centre in the world, fix a point in the homogeneity of profane space, 'provide a place where we can inhabit the world, *and* they provide vertical connexion with the gods' (Game, 2001: 236, italics in original; see also Eliade 1959).

Spatial forms and architecture across religious traditions – poles, pillars, spires, temples, steeples, bell towers, and domes – give expression to this *connexion* to the cosmos. Within prison, however, such a cosmological *connexion* must be expressed in other forms, as security and safety measures supersede religious and spiritual furnishings and accommodations (Beckford, 2001; Waldram, 1997). It is a truism to say that space, spatial boundaries, and spatial segregation are extremely intense in prison. Indeed, the aphorism by John-Paul Sartre – 'Hell is other people' – applies particularly well to spaces of incarceration. The ability to establish at least some 'ownership' over space is endemic among inmates, especially with the extreme overcrowded conditions within prisons that have increased exponentially in the last decade in Western countries (Beckford, 2001; O'Reilly-Shaughnessy, 2001). So, the need for inmates to secure a sacred space to practise their religion or spirituality becomes significant in relation to spatiality.

Moreover, having an Indigenous sample majority demands that conceptualisations of carceral spaces must consider sweat lodges and the spiritual sacredness of such rituals. When asked about whether there was a space in prison where the other men could let their guard down and be more emotional (happier, sadder, etc.), the men talked of such spirituality and sacred spaces. For instance,

Jacob reported that equality between and acceptance of inmates was significant to the sacredness of the chapel and spirit grounds. Especially in terms of the latter:

> [The Elders] always ask you to leave your human ego at the gate. Once you enter the sacred grounds you [enter] as a humble child … Nobody's better, nobody's stronger, no one's better looking. We're all equal there. And the camaraderie there was really good, you know, there was no phony guys there. You could just be yourself there because, again, it's a place of … worship … They always bring different groups in, male, female, so you can sit there and yap … And then, you know, you have to go back to your cell after. But, you know, you left [the spirit grounds] in good spirits and if you didn't allow anybody to, to pop your bubble, you could carry that for a few days, you know what I mean. (Jacob)

Indeed, several men mentioned feelings of relaxation and contentment when entering either the chapel or the sweat lodge. By fostering this *connexion* to the sacred, organised religion and spirituality play an important role in assisting men coming to terms with their identities, their feelings, and their ontological selfhoods. For example, it is common for religious and/or spiritual inmates to develop exceptionally powerful attachments to things such as photographs, images, books, talismans, and religious artefacts (Beckford, 2001). Two men raised the importance of sacred objects in their narratives: Adrian, referring to the cross pendant he wears around his neck; and Jacob, regarding the Indigenous drum he had made for himself to give to his life partner once he was released:

> [W]henever I think of picking up the bottle I just hold my cross … and I just think of all the chances God gave me and I'm just very thankful and I just don't want to drink anymore. I just don't want to let him down. (Adrian)
>
> I learned how to make drums. And I figured out what a drum is about, you know. It's okay to learn how to make them but there's a purpose of why you make things, instead of just making them to look nice. So I made two drums. I have one I call the 'four direction' drum because it took me four times. After I made it, it didn't work so I'd take it apart, four times. That's why I call it the 'four direction' drum. And, I made it for my partner … [O]nce I learned to calm myself and not allow situations to bother me, I made the drum and it sounds beautiful, it sounds very nice. So I gave that to her when I got out. The second drum I made was for myself, and I made that one really nice … [Y]ou know, it's kind of funny, I had to go to jail *to find myself*, I had to go to jail *to find my culture*, and I had to go to jail *to find out what a real man is about*. Yeah, and that's, you know, I'm grateful for it. You know, some people think I'm crazy when I say that, 'I'm grateful that I went to jail', you know, because it made me realise who I was. (Jacob, italics added)

In order for Jacob to 'find' himself, the prison space was significant to such self-discovery and cultural recovery. Jacob emphasises a selfhood that not only takes into account conflict resolution and anger management, but roots his identity construction in Indigenous culture and its teachings. Although there are certain ceremonies and rituals that are taught by the Elders and Indigenous teachers, 'each and every prisoner takes something different from these stories and teachings, something that they themselves need for their own personal healing' (O'Reilly-Shaughnessy, 2001: 99). Conceptualising his own perspective of what defines a 'real man', Jacob was able to sensually disengage from the mundanity and pitfalls of prison life and forge a strong spiritual *connexion* within his inner space, thus fostering a positive self-development for himself upon release.

Organised religion and spirituality assists men in coming to terms with not only their identities, but also their feelings and ontological selfhoods. Spirituality, then, becomes 'a personal connexion to something else', a bridge towards phantasmagorical spaces (O'Reilly-Shaughnessy, 2001: 98). Inmates are more likely to display anger and hurt if the sacredness of these objects are defiled by other inmates or staff, when such objects are invested with exceptional power and value (Beckford, 2001). While the full extent of the role of religion and Indigenous spirituality in the men's lives is beyond the scope of this study, the men's narratives illustrate how the sacredness of both the objects and spaces within prison must be guarded and upheld in order to preserve the sanctity of the inmate's *connexion* to the sacred and the mobility between the sacred and the carceral space.

Therefore, once inmates travel to their mental spaces to escape reality, it is possible that these men '[draw] from the imaginary the means of transcending movement, to such an extent as to be able to retrace the road of life' (Bailley, 1993: 248). The movement between the sacred and the carceral space indicates that the mind can be the place of one's rebirth. This journey '[into] the mind ... provides access to a marvelous elsewhere', where one attains new knowledge of the self and revitalises the mind and body to endure the struggles of prison life (Bailley, 1993: 248). Ultimately, movement and place are essential components to the existential travel between the real and the imaginary, and by forging a strong spiritual *connexion* to the sacred, Indigenous inmates form positive self-conceptions from cultural teachings.

Conclusion

New comprehensions of space are witnessed when we situate prison into the literature of imaginative mobilities and carceral geographies. Such comprehensions might be called what Thrift refers to as 'movement-space': 'folded and animate because everything can be framed in perpetual movement,' constituting the prison as a 'perpetually mobile space ... open-ended rather than enclosed' (2005: 592). Indeed, this chapter has demonstrated how the human mind itself seems to symbolically resemble the 'bridge' and 'door' argument Simmel suggests about space. Consciousness forms a bridge to imagination that the inmate transcends, providing a 'wonderful feeling of floating for a moment between heaven and earth' (Simmel, 1994: 8). Simultaneously, as a door, the imaginative geography of the mind flows life 'out of the door', from the limitation of 'isolated separate existence' within the prison environment into 'the limitlessness of all possible directions' (Simmel, 1994: 8).

This chapter has explored how the *connexion* between spaces of incarceration and one's ability to travel to the inner sanctums of the mind contributes to the growing carceral geographical scholarship. Such bodily and mental activities provide inmates with vehicles for self-expression and mental and pseudo-physical freedom. The cultivation of the body and inner psyche becomes simultaneously a source of personal liberation and social control. This migration to find inner peace and tranquility within a prison culture of (in)security suggests that not only does the physical prison environment play a role in the well-being of the inmate, but that the private spaces of the mind are key to understanding how inmates survive within carceral spaces. Forced to endure the boredom, mundanity, uncertainty, and (in)security of the prison culture, imaginative inner spaces fosters creativity and entertainment for the men trapped behind the prison walls.

The overrepresentation of Indigenous men within the Canadian corrections system remains fraught with ongoing challenges that cannot be remedied easily (Comack, 2008; Gacek, 2015; Waldram, 1997). However, former-inmate narratives are significant to demonstrate how researchers should reconsider the issues of 'doing time'. Such narratives are not so much tales of men battling against the prison industrial complex. Rather, these men's triumphs say something about the potential of imagination and movement, their ability to 'sustain sanity in an insane [carceral] place', and their resistance against the impact of incarceration (Kupers, 2001: 62). Examining the struggles that inmates face while incarcerated provides a conceptual entry point to the greater issues surrounding Canadian prisons and spatiality.

Acknowledgements

Generous financial support for the project was provided by the Social Sciences and Humanities Research Council of Canada through the Manitoba Research Alliance grant, 'Partnering for Change: Community-Based Solutions for Aboriginal and Inner-City Poverty'.

Notes

1 The average age of the men in the sample was 30.3 years old (the lowest age was 20, while the highest age was 47 years old). Recollections of time served in correctional institutions were approximations at best, as the men found it difficult to exactly remember the range of time spent for each prison sentence they had received or the total number of months or years spent within a particular correctional institution. The incorporation of the men's narratives into the study, coupled with an understanding of what the narratives are trying to say about carceral experiences, highlights the significance of acknowledging the human aspects that exist within data collection and analysis.
2 In particular, there are four correctional institutions Manitoba that remand males into custody: Winnipeg Remand Centre is a pretrial detention centre located in downtown Winnipeg. This centre houses people waiting for court decisions on their charges or placement in correctional institutions; Milner Ridge Correctional Centre is a medium-security institution located in the Agassiz Provincial Forest; Headingley Correctional Centre is a minimum-, medium-, and maximum-security institution located in Headingley,

Manitoba; and Stony Mountain Correctional Centre is a federal institution that offers minimum, medium and maximum security and is located in Stony Mountain, Manitoba.

3 Pseudonyms were assigned to respondents in order to ensure their confidentiality.

4 A card game resembling rummy, using two packs of cards. It is usually played by two pairs of partners, and the aim is to collect sets (or melds) of cards.

References

Abu El-Haj N (2013) Edward Said and the political present. In: Netton IR (ed) *Orientalism Revisited: Art, Land and Voyage*. Abingdon: Routledge, 55–86.

Baer LD and Ravneberg B (2008) The outside and inside in Norwegian and English prisons *Geografiska Annaler: Series B, Human Geography* 90(2): 205–216.

Bailley AS (1993) Spatial imaginary and geography: A plea for the geography of representations. *GeoJournal* 31(3): 247–250.

Beckford JA (2001) Doing time: Space, time religious diversity and the sacred in prisons. *International Review of Sociology: Revue Internationale de Sociologie* 11(3): 371–382.

Bowker LH (1980) *Prison Victimization*. New York, NJ: Elsevier.

Comack E (2008) *Out There/In Here: Masculinity, Violence and Prisoning*. Winnipeg: Fernwood Publishing.

Crewe B, Warr J, Bennett P and Smith A (2014) The emotional geography of prison life. *Theoretical Criminology* 18(1): 56–74.

Edensor T (2013) Reconnecting with darkness: Gloomy landscapes, lightless places. *Social & Cultural Geography* 14(4): 446–465.

Edensor T (2015a) The gloomy city: Rethinking the relationship between light and dark. *Urban Studies* 52(3): 422–438.

Edensor T (2015b) Introduction to geographies of darkness. *Cultural Geographies* 22(4): 559–565.

Eliade M (1959) *The Sacred and The Profane: The Nature of Religion*. London: Harcourt Brace & Company.

Farrington K (1992) The modern prison as total institution? Public perception versus objective reality. *Criminal Delinquency* 38(1): 6–26.

Finanne M and McGuire J (2001) The uses of punishment and exile: Aborigines in colonial Australia. *Punishment & Society* 3(2): 279–298.

Flyvbjerg B (2006) Five misunderstandings about case-study research. *Qualitative Inquiry* 12(2): 219–245.

Foote KE (2003) *Shadowed Ground: America's Landscapes of Violence and Tragedy*. Austin, TX: University of Austin Press.

Foucault M (1977) *Discipline and Punish: The Birth of the Prison*. Harmondsworth: Penguin.

Foucault M (1998) *Aesthetics, Method, and Epistemology*. New York, NY: The New Press.

Frank MC (2009) Imaginative geography as a travelling concept. *European Journal of English Studies* 13(1): 61–77.

Gacek J (2015) The impact of 'life' behind bars: Understanding space, impression management, and masculinity through former inmate narratives. Unpublished Masters thesis, University of Manitoba.

Game A (2001) Belonging: Experience in sacred time and space. In: May J and Thrift N (eds) *TimeSpace: Geographies of Temporality*. London: Routledge, 226–239.

Hancock P and Jewkes Y (2011) Architectures of incarceration: The spatial pains of imprisonment. *Punishment & Society* 13(5): 611–629.

Hannam K, Sheller M and Urry J (2006) Editorial: Mobilities, immobilities, and moorings. *Mobilities* 1(1): 1–22.

Hogg R (2001) Penality and modes of regulation of indigenous peoples in Australia. *Punishment & Society* 3(3): 355–379.

Jewkes Y (2013) On carceral space and agency. In: Moran D, Gill N and Conlon D (eds) *Carceral Spaces: Mobility and Agency in Imprisonment and Migrant Detention*. Farnham: Ashgate, 127–131.

Kerste A and Gilardi R (2003) The barren landscape: Reading US corporate architecture. In: Carr A and Hancock P (eds) *Art and Aesthetics at Work*. Basingstoke: Palgrave Macmillan, 138–154.

Kobayashi A and Peake L (2000) Racism out of place: Thoughts on whiteness and antiracist geography in the new millennium. *Annals of the Association of American Geographers* 90(2): 392–403.

Kupers TA (2001) Doing time, doing masculinity: Sports and prison. In: Sabo D, Kupers TA and London W (eds) *Prison Masculinities*. Philadelphia, PA: Temple University Press, 61–66.

Larsen J, Axhausen KW and Urry J (2006) Geographies of social networks: Meetings, travel and communications. *Mobilities* 1(2): 261–283.

McKegney S (2014) *Masculindians: Conversations about Indigenous Manhood*. East Lansing, MI: Michigan State University Press.

Milhaud O and Moran D (2013) Penal space and privacy in French and Russian Prisons. In: Moran D, Gill N and Conlon D (eds) *Carceral Spaces: Mobility and Agency in Imprisonment and Migrant Detention*. Farnham: Ashgate, 167–182.

Moran D (2012) 'Doing time' in carceral space: TimeSpace and carceral geography. *Geografiska Annaler, Series B* 94(4): 305–316.

Moran D (2013a) Between outside and inside? Prison visiting rooms as liminal carceral spaces. *GeoJournal* 78(2): 339–351.

Moran D (2013b) Carceral geography and the spatialities of prison visiting: Visitation, recidivism, and hyperincarceration. *Environment and Planning D: Society and Space* 31(1): 174–190.

Moran D (2015) *Carceral Geography: Spaces and Practices of Incarceration*. Farnham: Ashgate.

Moran D, Piacentini L and Pallot J (2013) Liminal transcarceral spaces: Prison transportation for women in the Russian Federation. In: Moran D, Gill N and Conlon D (eds) *Carceral Spaces: Mobility and Agency in Imprisonment and Migrant Detention*. Farnham: Ashgate, 109–124.

O'Reilly-Shaughnessy P (2001) Friction within the machine: Aboriginal prisoners behind the Wall. Unpublished Masters thesis, University of Manitoba.

Price PL (2010) At the crossroads: Critical race theory and critical geographies of race. *Progress in Human Geography* 34(2): 147–174.

Simmel G (1994) Bridge and door. *Theory, Culture & Society* 11(1): 5–10.

Simmel G (1997) The sociology of space. In: Frisby D and Featherstone M (eds) *Simmel on Culture*. London: Sage, 137–185.

Tataro C (2006) Watered down: Partial implementation of the new generation jail philosophy. *The Prison Journal* 86(3): 284–300.

Thrift N (2005) But malice aforethought: Cities and the natural history of hatred. *Transactions of the Institute of British Geographers* 30(4): 133–150.

Urry J (2007) *Mobilities*. Cambridge: Polity Press.

Urry J (2008) Moving on the mobility turn. In: Canzler W, Kaufman V and Kesselring S (eds) *Tracing Mobilities: Towards a Cosmopolitan Perspective*. Farnham: Ashgate, 13–35.

Urry J (2010) Mobile Sociology. *The British Journal of Sociology*: 347–366.

Urry J (2012) Social networks, mobile lives and social inequalities. *Journal of Transport Geography* 21: 24–30.

Waldram JB (1997) *The Way of the Pipe: Aboriginal Spirituality and Symbolic Healing in Canadian Prisons*. Peterborough: Broadview Press.

Williams SJ (2005) *Sleep and Society: Sociological Ventures into the (Un)known*. Abingdon: Routledge.

Woolford A and Gacek J (2016) Genocidal carcerality and Indian residential schools in Canada. *Punishment & Society* 00: 1–20.

7 Spreading the word

The dissemination of the American convict code, 1919–1940

Alex Tepperman

Introduction

In tracing the lineage of the American Prisoner's Rights Movement, historians of penology most often look to a series of landmark legal decisions that, throughout the 1960s, established a wide variety of due process rights for both convicts and the accused on state and federal levels. In the rare instances in which scholars try to *locate* the social origins of this legal revolution, they most often point to a series of inmate riots that gripped more than fifty maximum-security prisons over an eighteen-month stretch between 1952 and 1953 (Hawkins, 1976; Pallas and Berber, 1973). Motivated by the need for better food and improved facilities, thousands of inmates were mobilised around the United States, setting fires, taking hostages, and demanding national news coverage of their appeals, bringing the central concerns of what would later become the formal Prisoners' Rights Movement into the public consciousness. While discussions of these riots are invaluable for framing the grassroots origins of later legal challenges, penal historians have yet to undertake a systematic search for the movement's earliest roots, much less a study of how inmates' ideas and aspirations spread throughout the country, despite the existence of excellent scholarship on the early histories of other major civil rights movements (Chauncey, 1994; Kelley, 1994) which might serve as useful models for understanding the ways in which the civil rights ethos was spread and disseminated from region to region and, ultimately, became the subject of a national dialogue.

Marie Gottschalk's *The Prison and the Gallows* (2006) is the best approximation of what a *long dureé* history of the Prisoners' Rights Movement might look like. Framing organised convict rebellion as an effect of 'the dramatic change in the US prison population and the nature of prison protests between the 1930s and 1970s (Gottschalk, 2006: 169), she traces the origins of the movement back to the three decades preceding its legal incarnation, thereby facilitating a broad reconsideration of how scholars might periodise organised prisoner rebellion. This chapter builds upon Gottschalk's work by questioning her contention that, prior to the 1930s, 'people who were imprisoned lacked significant internal organizations to sustain their *mobilizations*' (2006: 170, emphasis added), suggesting instead that an exceptional amount of inmate mobility between prison systems during the 1920s and 1930s facilitated powerful, though informal, forms of inmate organisation that made mass resistance possible through exchange and dissemination of ideas.

The historical record seems to suggest that the antecedents of a collective, national prisoner identity stem from the interwar period (1919–1940). During that two-decade stretch, the United States underwent one of its most radical periods of state and federal penological transformations. Due to an unprecedented increase in total prison populations and a major geographical redistribution of both free persons and inmates, prisoners found themselves able to develop and disseminate a recognisable, proto-national convict culture that mid-century social scientists readily observed in nearly identical forms throughout the country. Neither the existence nor the rapid movement of this collectivist ethos – consistently identified by ethnographers and inmates as the 'convict code' – means that, at any point, the country's inmates consciously saw themselves as working in common cause. These large, newly mobile convict populations simply lived within a social and political context that inevitably led to constant intermingling in overcrowded prisons between peoples from a broad swath of geographic regions and demographic profiles. This penal model, in turn, necessitated cultural negotiation among diverse prisoner populations who carried their lived experiences with them between various prison systems throughout the country, with the resulting 'convict code' laying out an unofficial moral compass for inmates.

In exploring dramatic changes in public policy and penal architecture, as well as the increasing racial and cultural heterogeneity of state and federal prison communities, this chapter will show that the mobilities of both free and interred populations encouraged meaningful segmental bonding among inmates and, ultimately, a cross-cultural code of convict mores to be *mobilised*. Convicts, having no historical distance from their own lives, would not have contextualised the broader meanings and long-term significance of their regular social exchanges as they happened and, for that reason, there exists no clear path, either through archival records or oral histories, to tracking the gradual movement of inmates' ideas, habits, and mores across space and time. For that reason, this chapter offers only one of many potential perspectives on the mobilisation and national distribution of a 'convict code,' extrapolating from two primary case studies – the Leavenworth federal penitentiary in Kansas and the New York state prison at Auburn – and the national movement of interwar prisoners and their cultures between the wars. Inmate case files from these institutions provide a glimpse into the prisoner experience on a national level, as Leavenworth and Auburn were both enormous, self-contained fortress-style prisons, retrofitted for the extraordinary interwar expansion of convict populations, mirroring many of the challenges facing maximum-security penitentiaries and their residents throughout the country during the 1920s and 1930s.

In addressing the informal movement of inmate culture through the mass movement of prisoners nationwide, this chapter will also attempt to further the use of historiographical methods and perspectives in mobilities studies more broadly.

While mobilities scholars have regularly engaged with questions regarding origination and long-term development, drawing on the works of Charles Tilly (1984) and Norbert Elias (1978), mobilities literature may benefit from more consistently and deliberately considering historiographical perspectives on periodization and narrative. In particular, in discussing the anti-crime moral panic of the 1920s and the sustained prison-building schemes and mass migrations of the interwar era, this chapter will place social and cultural history perspectives in conversation with related mobility literatures.

How new prison populations shaped the convict code

Politicians and large segments of the public believed that the end of the First World War brought with it a national crime wave (Adler, 2015: 34–35), and this subsequent crime panic prompted experimentations in new, often highly punitive lawmaking. While it may seem anachronistic to apply the highly presentist idea of 'mass imprisonment' to the early twentieth century, the United States grappled with major changes in the makeup of its convict populations and unprecedented levels of penal severity throughout the interwar period.

The punitive turn seems to have emerged largely in response to generalised anxieties about broadscale changes in American life. The 1920 federal census reported that, for the first time, a majority of Americans lived in urban areas. This led many national leaders to publicly lament what they saw as the passing of traditional, rural American life (Gusfield, 1963). Protestant, Anglo-Saxon citizens looked on with concern as fast-growing industrial hubs and urban polyglots drew millions of Southern blacks and Eastern and Southern European émigrés to the Northeast and Midwest. Whether rapid urbanisation and a major re-alignment in the racial makeup of the country caused an actual rise in crime is unclear, but it unquestionably changed the face of American prison populations.

In response to these macro-sociological changes, state and federal legal systems dealt with highly mobile, underemployed immigrant and black populations in an increasingly draconian fashion. On the state level, a number of densely populated regions, most notably California and Michigan, aped New York's passage of the hyper-punitive Habitual Criminal Act. Enacted in 1926, the New York State Crime Commission put forth a series of measures imposing significant minimum penalties for second felony offences and mandatory life sentences for those offenders convicted of a fourth felony. These measures would serve as the progenitors of the 'Three Strikes' policies of the 1990s.[1] While the federal government did not adapt similar mandatory minimum-style sentencing, it too passed a spate of sweeping felony laws, most notably the 1919 Volstead Act, which outlawed the sale and consumption of liquor throughout the United States. Accordingly, federal prison populations also exploded during the interwar era.

Public anxiety surrounding crime and the presence of large numbers of racial 'others' in quickly urbanising industrial regions fed an increase in per capita prison populations throughout the interwar period that the nation would not approach again until the 1980s. As demonstrated in Table 7.1, national rates of

Table 7.1 National incarceration rates (per 100,000), 1910–1950

System	1910	1923	1930	1940	1950
National	75	74	98	125	118
Federal	2	4	11	15	11
New York	78	58	65	114	107

incarceration dropped slightly between 1910 and 1923, only to rise significantly by 1930, increasing a total of 68.9 per cent between 1923 and 1940.

Rising federal prison populations were particularly pronounced, as per capita inmate populations doubled between 1910 and 1923 and quadrupled by 1940. As of 1920, the federal Department of Justice only held 4 per cent of the nation's prisoners and, for most of its history, either kept offenders in one of three maximum-security institutions or simply paid state and local facilities to hold less-troublesome prisoners. This quickly became untenable, however, as the federal system held almost 10 per cent of all inmates in the United States by 1940. Accordingly, the federal government embarked on a massive expansion of its prison system throughout the 1930s, constructing new high-capacity men's penitentiaries at El Reno, Oklahoma, and Graterford, Pennsylvania, along with a large men's reformatory at Chillicothe, Ohio.

To accommodate extraordinary increases in convict populations, prison administrators entertained both ambitious penitentiary building projects and a new willingness to overcrowd existing institutions. The 'Big House' style of prison, a form of penitentiary designed to be so large that it had sufficient space to simulate a free society behind bars, quickly became the interwar model for maximum-security internment (Rotman, 1995). Politicians, journalists, and academics frequently compared the early-twentieth-century Big House penitentiaries to 'nations' and 'cities,' looking to such iconic examples as Michigan's sixty-four acre Jackson State Prison – completed in 1924 – and its 'water and power, telephone system, industries, garage, laundry, hospital, school, theater, library, newspaper, [and] band' (Cox, 2009: 7) as a model for the future. This scale of imprisonment was increasingly normalised over the interwar period, however, as even Big House institutions were, by the end of the 1930s, wildly overcrowded at an average of around 2,500 inmates. Such rampant stuffing provided a direct challenge to the conventional wisdom among mid-century criminologists that holding more than 1,000 to 1,200 inmates in any institution, regardless of its size, would critically undermine the administration's ability to manage their wards (Barnes and Teeters, 1959: 484-485).

The very 'bigness' of the Big House seems to have seduced prison administrators into a misplaced overreliance on that form of imprisonment. While the national prisoner population increased 30.7 per cent between 1910 and 1930, the population of the 34 prisons throughout the nation that held 1,000 or more inmates increased by an average of 76.9 per cent between 1915 and 1929.[2] As Figure 7.1 indicates, the nation's largest institutions were growing more than twice as quickly as the national prison population on the whole by the eve of the Great Depression, housing an average of 937 more prisoners than they held before the start of the war.

The inevitable over-packing of inmates turned penitentiaries into warehouses for social undesirables. Michigan may have provided the best example of this tendency, as its Ionia Correctional and Jackson State each tripled in size throughout the interwar period, as did the federal penitentiaries at Leavenworth and Atlanta. The Ohio State Penitentiary at Columbus and California's San Quentin State Prison similarly served as powerful symbols of institutional overextension, topping out at over 4,000 residents each. Table 7.2 indicates how crippling overcrowding was in the nation's very largest institutions.

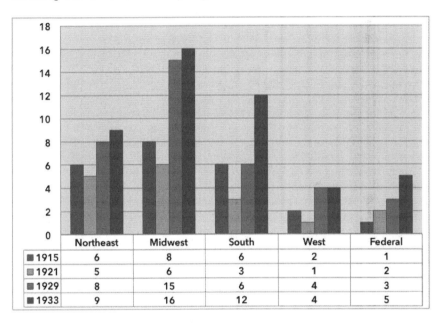

	Northeast	Midwest	South	West	Federal
■ 1915	6	8	6	2	1
▨ 1921	5	6	3	1	2
▨ 1929	8	15	6	4	3
■ 1933	9	16	12	4	5

Figure 7.1. Number of penitentiaries with 1000+ inmates, by region, 1915–1933.
(Produced by the author.)

Table 7.2. Big house prisons, 1929, by population, rank, capacity, and capacity filled

Institution	System	1929 Population	National Rank	Capacity	Capacity %
San Quentin	California	4638	1	2844	163
State Penitentiary	Ohio	4182	2	1500	279
Leavenworth	Federal	3561	5	1560	228
Stateville	Illinois	3492	6	2152	162

The understandable difficulties administrators had in controlling inmates within Big House prisons were augmented by the constant movement of populations between federal-, state-, and county-level institutions that left the composition of penal populations in flux. As such, the interwar period saw extreme levels of prisoner mobility and, in both Auburn and Leavenworth, the number of inmates who arrived with personal experiences of unaffiliated prison systems – either having served prison time in a different state or, in Auburn's case, in the federal system – soared during the second half of the interwar era. Between 1929 and 1940, 13.3 per cent of Auburn's new inmates and 27.6 per cent of Leavenworth's incoming inmates had previously served in unaffiliated systems, a remarkable rise from the respective 5.7 per cent and 12.7 per cent rates of the first half of the interwar period. This trend was especially pronounced for Leavenworth, as the number of Leavenworth recidivists who had served in two or more non-federal prison systems before entering federal custody rose from 28.4 per cent during the first half of the interwar period (1919–1928) to 64.2 per cent in the second half (1929–1940), with those having already served in three or more unaffiliated prison systems rising to 14.3 per cent. Put another way, almost one in seven incoming Depression-era Leavenworth inmates experienced life in at least four distinct prison systems.[3]

While the dynamic movement of inmates around the country was surely a disorienting experience for many, convicts of the time may also have been more adept at maintaining and adapting their value systems to new environments than is generally assumed. Southern blacks, one of the most visible and overrepresented groups in interwar penitentiaries, moved to the North and West in the hundreds of thousands as part of the Great Migration of the 1910s and 1920s, carrying the experiences of Southern life with them. Other ubiquitous prison populations included World War One veterans, homeless drifters, hoboes, and significant immigrant populations in states like New York and California. Each of these highly mobile groups, through their lifestyles as transient people, developed senses of social solidarity and proved themselves capable of adapting to life and maintaining their cultures in alien environments. In circulating around the country and the world as free peoples, a large proportion of interwar convicts embodied a sort of 'double mobility' by also moving between prison systems. Ultimately, the critical mass of inmates, the constant introduction of new inmates from diverse penal backgrounds, and the relatively worldly nature of American inmates were all crucial to the creation of a national prisoner culture and subsequent 'convict code'.

Circulating the prison body

The fact that recidivists were, by the interwar period, coming from increasingly diverse institutional backgrounds meant thousands of inmates from a wide array of penal cultures were regularly cycled throughout the country's largest penitentiaries. This effectively created a cross-national pipeline of repeat offenders who either consciously or unconsciously exported, and disseminated fragments of their local prison cultures to new institutional homes. One can see the faint outline of what was surely a deeply complex, tangled mass of communications between disparate inmate groups by looking at which prison systems served as feeders for others.

As a means of adding additional texture to the national picture, this section will focus not only upon Leavenworth and Auburn, but also on Louisiana's Angola Prison Farm. Angola provides a valuable counterpoint to Auburn and Leavenworth, both of which were flagship maximum-security institutions in relatively progressive penal systems, each serving as models for national correctional policies throughout the twentieth century. Held out as America's most iconic prison farm (Schrift, 2004; Berger, 2014: 173), Angola has remained an important emblem of the racially segregated Deep South's early antipathy to modern Big House corrections and provides some sense of how prison populations moved through institutions that were mostly cut off from the broader, national prison-building projects of the 1920s.

Geographical proximity was, perhaps unsurprisingly, the most telling indicator of which system New York's interwar era recidivists had served in previously, as the adjacent states of Pennsylvania, Ohio, and New Jersey were the most commonly represented within Auburn's prisoner records. Alternately, for Leavenworth, the fact that it was not anchored to a single geographical space meant the federal system was a uniquely important force in the nationalising of prisoner culture. Not only was Leavenworth home to significant numbers of inmates from largely rural, southwestern states like Texas, Oklahoma, Missouri, and Louisiana, but former federal inmates were present in significant numbers in major industrial states like California, New York, Michigan, and Illinois, which stretched from coast to coast. As Tables 7.3 and 7.4 indicate, Auburn and Leavenworth's records, when taken together, show significant connections between eleven major prison systems that, collectively, held more than three-fifths of America's total prison population.

Angola provides a distinct counterpoint, as Louisiana was more clearly guided by a provincial white supremacist ethos that closely represented the power structure and race politics of the Deep South but not necessarily of the nation as a whole. In direct contrast to Auburn and Leavenworth, where the black inmates

Table 7.3. Prison systems in which Auburn inmates previously served, and percentage of national prison population each system represents, 1919–1940

State	National Prison Population %	Representation in Auburn, per 1000
New York	6.8	∞
New Jersey	2.4	13
Ohio	7.3	11.5
Pennsylvania	4.4	18

Table 7.4. Prison systems in which Leavenworth inmates previously served, and
percentage of national prison population each system represents, 1919–1940

State	National Prison Population %	Representation in Leavenworth, per 1000
Federal	10.8	∞
Missouri	3.3	13.2
Michigan	5.9	17.6
Oklahoma	3.1	17.1
Illinois	6.6	14.2
Texas	4.2	15.1
California	5.9	15.6

normally made up between 10 and 15 per cent of the total population – slightly below the interwar national average of 21 per cent – African-Americans made up 70 per cent of Angola's inmates, despite the state being majority-white. The hyper-racialised nature of imprisonment in Louisiana is also noticeable in the fact that the only state prison systems represented in significant numbers among Angola inmates were neighbouring, and similarly segregationist, Texas and Arkansas. The sparse representation of other prison systems within the Angola population was likely due, at least in part, to the difficulties Southern blacks faced in moving freely about the South, as 70 per cent of Angola inmates came from Louisiana, whereas under half of inmates at Auburn were born in-state.

These differences are typical of the North-South divide throughout the early twentieth century. The 'Deep,' or 'Lower,' South was the slowest region to adopt the Big House prison style, hanging on to the older, highly racialised, and deeply authoritarian system of convict leasing. The leasing system had emerged from the unique political environment of the post-Reconstruction New South, as the fall of America's slave society in 1865 and the removal of federal troops from formerly Confederate states in 1877 led Southern legislatures to develop a form of neo-slavery in which black convicts, rather than serve prison sentences, were made to work in chained gangs, often under the supervision of private contractors. With black Americans frequently subject to harassment from local law enforcement, the convict lease system allowed Southern governments to practice crypto-slavery well into the twentieth century.

The South did not fully end the highly profitable practice of forced prison labour until Alabama abolished its leasing system in 1928 (Oshinsky, 1997). The end of convict leasing seemed to signify a fundamental change in Southern attitudes toward corrections. Prior to 1929, only four of the nation's

34 1,000-inmate-or-greater prisons were located in the South, and even they were located in the regional fringes of Tennessee, Virginia, Kentucky, and Oklahoma. Between 1929 and 1933, however, another eight Southern prisons passed the 1,000 inmate threshold and, by the late 1930s, the Deep South had mostly adopted mainstream penological practices, replacing traditionally agrarian or forced labour–based punishments with more contemporary machineries of power that tied into the interstate systems of convict exchange found in Auburn and Leavenworth.[4]

The South's role as a regional bulwark of white supremacy and a significant barrier to the ability of free black Americans to move about the country played its own indirect role in the development of a national prisoner culture. The disproportionate presence of blacks in the Deep South, and their relative unlikeliness to be present in larger, non-Southern prisons suggests that the origins of the convict code were disproportionately propagated by white non-Southerners and would not necessarily represent the specific needs and concerns of black inmates going forward. This would become a significant problem in the late 1960s, as the presence of new urban gang cultures, Nation of Islam practitioners, and a Black Power movement steeped in revolutionary Marxist discourse would make interracial political cooperation among American inmates far less plausible than had been the case in the white-dominated prisons of earlier generations.

The Convict Code

Defined by their oversized architecture and double-capacity populations, multitudes of hidden corners and unwatched interactions behind bars made Big House prisons largely ungovernable. Certainly, wardens could never hope to maintain 'Separate System'-style control over their wards,[5] though that remained the ideal condition. Accordingly, administrations throughout the nation abandoned older Progressive Era ideas of reform, focusing instead on simply 'managing' prison populations.[6] In their understaffed institutions, wardens and guards could only hope to keep their enormous populations sated and the prison's formal mechanisms running smoothly. Administrators lacked the space, funds, and manpower to keep a strict eye on inmates in areas like dining halls and recreation yards. In turn, those became important areas of prisoner contact where inmates could chat and gossip, pass notes, fight, play, and otherwise pass time. In turn, a newly empowered 'society of captives' flourished (Clemmer, 1940; Sykes, 1956).

Indeed, one of the more widely disseminated elements of Big House prison culture was a clearly articulated set of principles and practices that inmates came to call the 'convict code'. The code set out a broad social contract for prisoners, establishing a variety of ideal behaviours that inmates of widely varying socio-economic and racial backgrounds, as well as sentence lengths, could agree to. Passed along orally until the 1930s, a variety of academics, journalists, and prison memoirists eventually brought elements of the code to the public. By the eve of the Second World War, multiple Hollywood films portrayed the code to public audiences, including inmates and future inmates themselves, thereby further facilitating the code's movement around the country.

Once adopted by the highly mobile prison populations of the interwar era, the code spread quickly. A content study of 20 prison ethnographies published between 1927 and 1965 provides a sense, if only an impressionistic one, of just how quick that process was. Of ten prominent studies of prison argot published between 1927 and 1953 – a set of works that cover seven different states and regions – only Hans Reimer's trailblazing 1936 study of the Kansas State Penitentiary at Lansing mentions a 'code' in any iteration. Of ten surveyed works covering the years 1954 to 1965 and areas ranging geographically from New Jersey to Hawaii, on the other hand, nine make explicit mention of an 'inmate code', 'prisoner code', or simply '*the* code'. While these findings are not necessarily indicative of national trends, they are at least suggestive of a broad, long-term movement of ideas across the country's penal landscape.[7]

From the perspective of mid-century American sociologists, the convict code seemed to have been imported from marginalised communities in the free world. Erving Goffman noted this, finding language and modes of coping overlapping among the 'human junk' present in hobo jungles, skid rows, and prisons (1952). Likewise, Gresham Sykes, in studying New Jersey State Prison, found that many of the ideas he was privy to could be traced back to 'prostitutes, tramps, beggars, [and] carnival workers' (1958: 84). Gresham Sykes and Sheldon Messinger (1960) summed up the code succinctly: Do not cooperate with the administration and do not victimise your fellow convicts. Irwin and Cressey echoed the same ideas more colloquially, describing the motivations as 'don't inform on or exploit another inmate, don't lose your head, be weak, or be a sucker' (1962: 145).

The code was steeped in the same concepts of group loyalty as the earlier thief's code, hobo code, and carnie code (Anderson, 1923; Easto and Truzzi, 1973), all of which aimed to set out guidelines for right or honourable behaviour among members of common social 'out groups'. Howard Becker (1963) referred to such measures as 'colleague codes', characterising them as a stated desire on the part of social outsiders to close ranks in the face of opposition from alien populations. The most important pillars of such codes involved the desire to maintain private spaces, to exclude others, to settle disputes without input from government or corporate agencies, to maintain secondary economies without relying on state-issued currency, and to withhold criticising others in one's same caste while in the presence of non-group members (Becker, 1963: 86).

Nationalised convict culture moved slowly and organically, emerging from a process of cultural syncretism within prison walls, as heterogeneous local and national inmate cultures merged and negotiated a shared terrain within rapidly expanding interwar Big Houses. The prison was, after all, a liminal area that provided space for what the historian Richard White (1991) coined a cultural 'middle ground' – a region in which disparate cultural groups had little choice but to negotiate mutually understandable notions of acceptable behaviour – among distinct prison populations. There, prisoners might chat or gossip, casually sharing their thoughts, language, and values in the recreation yard or during a meal. They might also, in a less-familial mood, establish or disseminate informal rules of conduct and punishment, laying out locally recognized parameters of acceptable or unacceptable behaviour for new inmates. Whether these cultural exchanges were steeped in friendship or enmity, however, they were rarely recorded for posterity, a regrettable result of the transient, ephemeral nature of much of prison life. Prisoners simply established and strengthened these cultures one conversation at a time.

The slow movement and negotiation of a shared convict culture over years suggests that the wending and winding of these ideas across the nation was a creative, not destructive, process. While Svensson and Svensson (cited in Moran, 2015: 79) compellingly argued that the cycling of inmates throughout a variety of systems and across a geographical spaces can serve to disrupt prisoners' cultural rules and rituals, the inverse may also be true. The constant reshuffling of inmates and their provincial penal cultures throughout state and national prison systems may have homogenised and strengthened elements of the national inmate population, spreading once-regional or local norms across the country. This was accomplished by the creation of what Cresswell calls 'mobility constellations', those being 'historically and geographically specific formations' (Cresswell, 2010: 17) borne of unique cultural circumstances. In American prisons, inmate-developed cultures were often well-defined on the local level while also adaptable enough to evolve to the needs of similar-but-distinct prison cultures around the country. These networks allowed American prisoner culture to be both local and national, regionally diverse but cosmologically aligned.

While the code was ubiquitous, clearly articulated, and broadly accepted, however, inmates never practiced its tenets universally. Convict 'ratting', or informing on other convicts in exchange for preferential treatment, was not only frequent but, in many cases, habitual. Noting a common betrayal of colleague codes within total institutions, Goffman argued that, while 'there is little group loyalty in total institutions', there is still an 'expectation that group loyalty should prevail' (Goffman, 1961: 61). The ideology's importance laid not in its execution, but in inmate populations' broad-scale acceptance of its general necessity and moral *justness*. Prisoners, administrators, and academics all understood the code was aspirational, not functional.

The work of the sociologists Norbert Elias and Eric Dunning (1986) posits some insight into the value the convict code had as a model for shaping inmate life. Elias and Dunning's use of segmental bonding to describe the social processes by which disempowered working-class populations create and encourage non-bourgeois countercultures provides a model for interpreting why diverse prison populations might gravitate toward the formation of a shared culture. In their investigation of British football hooliganism, the theorists also found a constellation of shared social characteristics among habitual rioters, including extreme poverty, high susceptibility to unemployment, low levels of formal education, 'intense feelings of attachment to narrowly-defined "we-groups" and correspondingly intense feelings of hostility towards narrowly-defined "they-groups"' (Elias and Dunning, 1986: 243), characteristics that mid-century penal ethnographers also found widely spread among convict populations. If one accepts the notion, then, that imprisonment is a deeply disorienting and distressing experience that creates anomie among inmates and, to some extent, levels out certain other 'real world' social divisions, one might see the creation of the convict code as a predictable form of segmental bonding, adopted by inmates as a means of coping with the common pains of internment.

Such discussions of shared values and bonding should absolutely not be read as a romantic portrayal of prison life or a belief in any sort of magnanimous 'live-and-let-live' détente among convicts, as that view would run contrary to prisoners' often-violent hyper-masculine 'home worlds' (Goffman, 1959), as well as to the extreme racism and cultural chauvinism of the interwar period. It does offer, how-ever, some insight into why an idealised social contract among inmates who had, throughout the interwar period, been forced together into increasingly crowded, under-watched prisons might develop. The circulation of a critical mass of con-victs allowed inmates an opportunity to create a contraculture based on the few common social ties that bounded inmates together.

Conclusion

Having examined the Prisoners' Rights Movement's prehistory, one gets some sense of how the growing discontent of a bloated interwar convict population and a web of interconnected penal systems allowed a steady procession of recidivists to carry their distinct prison cultures through Auburn, Leavenworth, and elsewhere. The thousands of inmates from around the country who tramped through Big House prisons turned what little free space existed in prisons into civic squares in which, either through coercion, persuasion, or active compromise, they came to negotiate common means of engagement. This change was neither instantane-ous nor uniform, but its broadest strokes reached all corners of the country. We see in the movement of prisoners and their values the beginnings of the convict consciousness that made later, more obvious acts of resistance possible.

One also finds in this chapter the potential methodological value history provides to the study of mobilities. After all, the interwar 'moment' that allowed for the development of an interwar prisoner culture by aligning a national moral crisis and crime panic, an unprecedented faith in mass imprisonment and prison overcrowding, and the overexpansion of federal and state prisons provided the necessary set of circumstances under which a negotiated convict culture might develop. In both the development and the recognition of the historical moment of the development, then, one can see the value of historical perspectives in the study of mobilities. Historiography is, after all, largely about understanding the importance of contingency, meaning-making, and implication. Certainly, a more deliberate consideration of historical perspectives could provide mobilities scholars with a fuller perspective of how to read movement as a historical process.

Notes

1 Also known as 'habitual offender laws', so-called Three Strikes policies are sentencing guidelines that provide mandatory, graduated levels of increasingly severe sentencing for repeat offenders. Metaphorically named after the 'three strikes and you're out' notion taken from baseball, these policies assert that any person sentenced to a third felony conviction would automatically receive a lifetime sentence, regardless of the offence or circumstances.
2 Unfortunately, only the Handbooks of American Prisons provide reliable inmate counts of individual prison populations for the early twentieth century, and even those counts mostly go back only to 1915.
3 By the interwar period, serving time in a wide variety of jails and penitentiaries around the country was a common experience for a variety of populations. Convicted resident aliens and migrant labourers were especially likely to find themselves systematically moved around and stored in all regions of the country, as the xenophobia of the 1910s and 1920s gave birth to a fully national, rail-based system of mass deportation. Unveiled in 1914, this complex network expanded to 181 distinct 'deportation parties' by 1931, featuring 19 'transcontinental specials' that reached from Seattle to Galveston, Texas to New York City. In the 1920s, this spiderweb of interconnected rail systems was moving thousands of non-citizen offenders across the country per year (Blue, 2015: 2).
4 In the Upper South, the Maryland House of Correction, Kentucky State Penitentiary, and the Lorton Reformatory in the District of Columbia all passed the 1,000-inmate mark by 1933, as did Alabama's Kilby Prison, Arkansas' Cummins Unit, the Florida State Prison at Raiford, and the South Carolina Penitentiary, all of which were located within the Deep South.
5 'Separate style' imprisonment refers to the eighteenth century model of imprisonment, inspired by Quaker notions of reflection and penitence, as well as by Bentham's notions of the panoptic gaze, first found international fame through its deployment at Pennsylvania's Eastern State Penitentiary.
6 The Progressive Era of American politics covered, approximately, the period from the late 1880s to the early 1910s. It worked off of a faith in the improvement of society through the social scientific 'helping professions' (city planning, social work, etc.) and held that government action could be used to uplift the underprivileged through both economic assistance and moral reform.
7 The works considered, all of which are included within the references, include Yenne (1927), Milburn (1931), Kuethe (1934), Hargan (1935), Reimer (1937), Hayner and Ash (1939), Clemmer (1940), Weinberg (1939), Hayner (1943), Haynes (1948), McCorkle and Korn (1954), Ohlin (1960), Caldwell (1956), Sykes (1958), Cloward (1960), Schrag (1959), McCleery (1960), Irwin and Cressey (1962), Garabedian (1963), and Jackson (1965).

References

Adler J (2015) Less crime, more punishment: Violence, race, and criminal justice in early twentieth-century America. *Journal of American History* 2: 34–46.
Anderson N (1923 [1998]) *On Hobos and Homelessness*. Chicago, IL: University of Chicago Press.
Barnes H and Teeters N (1959) *New Horizons in Criminology*. Saddle River, NJ: Prentice Hall.
Becker H (1963 [2008]) *Outsiders: Studies in the Sociology of Deviance*. New York, NY: Simon & Schuster.
Berger D (2014) *Captive Nation: Black Prison Organizing in the Civil Rights Era*. Chapel Hill, NC: University of North Carolina Press.

Blue E (2015) Strange passages: Carceral mobility and the liminal in the catastrophic history of American deportation. *National Identities* 17(2): 175–194.

Cahalan M (1986) *Historical Corrections Statistics in the United States, 1850–1984.* Rockville, MD: Westat.

Caldwell M (1956) Group dynamics in the prison community. *The Journal of Criminal Law, Criminology, and Police Science* 46(5): 648–657.

Chauncey G (1994) *Gay New York: Gender, Urban Culture, and the Making of the Gay Male World, 1890–1940.* New York, NY: BasicBooks.

Clemmer D (1940 [1958]). *The Prison Community.* New York, NY: Rinehart.

Cloward R (1960) Social control in prison. In: Hazelrigg L (ed) *Prison without Society: A Reader in Penology.* New York, NY: Doubleday, 78–112.

Cox S (2014) *The Big House: Image and Reality.* New Haven, CT: Yale University Press.

Cresswell T (2010) Towards a politics of mobility. *Environment and Planning D: Society and Space* 28(1): 17–31.

Easto P and Truzzi M (1973) Towards an ethnography of the carnival social system. *The Journal of Popular Culture* 6(3): 550–566.

Elias N (1978) *The Civilizing Process: The History of Manners.* New York, NY: Urizen.

Elias N and Dunning E (1986) *Quest for Excitement: Sport and Leisure in the Civilizing Process.* New York, NY: Blackwell.

Garabedian P (1963) Social roles and processes of socialization in the prison community. *Social Problems* 11(2): 139–152.

Goffman E (1952) On cooling the mark out: Some aspects of adaptation to failure. *Psychiatry: Journal of Interpersonal Relations* 15(4): 451–463.

Goffman E (1959) *The Presentation of Self in Everyday Life.* Garden City, NY: Doubleday.

Goffman E (1961) On the characteristics of total institutions. In: Cressey D (ed) *The Prison: Studies in Institutional Organization and Change.* New York, NY: Holt, Rinehart and Winston, 15–67.

Gottschalk M (2006) *The Prison and the Gallows: The Politics of Mass Incarceration in America.* New York, NY: Cambridge University Press.

Gusfield J (1963 [1986]) *Symbolic Crusade: Status Politics and the American Temperance Movement.* Urbana, IL: University of Illinois Press.

Hargan J (1935) The psychology of prison language. *The Journal of Abnormal and Social Psychology* 30(3): 359–365.

Hawkins G (1976) *The Prison: Policy and Practice.* Chicago, IL: University of Chicago Press.

Hayner N (1943) Washington state correctional institutions as communities. *Social Forces* 21(3): 316–322.

Hayner N and Ash E (1939) The prisoner community as a social group. *American Sociological Review* 4(3): 362–369.

Haynes FE (1948) Sociological study of the prison community. *Journal of Criminal Law and Criminology* 39(4): 432–440.

Irwin J and Cressey D (1962) Thieves, convicts and the inmate culture. *Social Problems* 10(2): 142–155.

Jackson B (1965) Prison folklore. *The Journal of American Folklore* 78(310): 317–329.

Kelley R (1994) *Race Rebels: Culture, Politics, and the Black Working Class.* New York, NY: Free Press.

Kuethe, JL (1934) Prison parlance. *American Speech* 9(1): 25–28.

McCleery R (1960) Communication patterns as bases of systems of authority and power. *Social Science Research Council* 15: 49–77.

McCorkle L and Korn R (1954) Resocialization within walls. *The Annals of The American Academy of Political and Social Science* 293: 88–98.

Milburn G (1931) Convicts' jargon. *American Speech* 6(6): 436–442.

Moran D (2015) *Carceral Geography: Spaces and Practices of Incarceration*. Farnham: Ashgate.

Ohlin L (1960) Conflicting interests in correctional objectives. *Social Science Research Council* 15: 111–129.

Oshinsky D (1997) *Worse Than Slavery: Parchman Farm and the Ordeal of Jim Crow Justice*. New York, NY: Free Press.

Pallas J and Berber R (1973) From riot to revolution. In: Olin Wright E (ed) *The Politics of Punishment: A Critical Analysis of Prisons in America*. New York, NY: Harper Colophon, 237–261.

Reimer H (1937) Socialization in the prison community. *Proceedings of the American Prison Association* 1937: 151–155.

Rotman E (1995) The failure of reform: United States, 1865–1965. In: Morris N and Rothman D (eds) *The Oxford History of the Prison: The Practice of Punishment in Western Society*. New York, NY: Oxford University Press, 151–177.

Schrift M (2004) The Angola Prison rodeo: Inmate cowboys and institutional tourism. *Ethnology* 43(4): 331–344.

Schrag C (1959) Social role, social position, and prison social structure. *Proceedings of the Eighty-Ninth Annual Congress of Correction of the American Correctional Association*. Washington, DC: American Correctional Association.

Sykes G (1956) Men, merchants, and toughs: A study of reactions to imprisonment. *Social Problems* 4(2): 130–138.

Sykes G (1958) *The Society of Captives: A Study of a Maximum Security Prison*. Princeton, NJ: Princeton University Press.

Sykes G and Messinger S (1960) The inmate social system. In: Cloward R (ed) *Theoretical Studies in Social Organization of the Prison*. New York, NY: Social Science Research Council, 5–20.

Tilly C (1984) *Big Structures, Large Processes, Huge Comparisons*. New York, NY: Russell Sage Foundation.

Weinberg SK (1939) Aspects of the prison's social structure. *American Journal of Sociology* 47(5): 717–726.

White R (1991) *The Middle Ground: Indians, Empires, and Republics in the Great Lakes Region, 1650–1815*. New York, NY: Cambridge University Press.

Yenne H (1927) Prison lingo. *American Speech* 26(6): 280–282.

8 Mobility and materialisation of the carceral

Examining immigration and immigration detention

Deirdre Conlon and Nancy Hiemstra

Introduction

In June 2015, Donald Trump announced his candidacy in the 2016 election for president of the United States. In the speech accompanying this announcement Trump offered several gross and sweeping misrepresentations of immigrants to the US, suggesting, 'They're bringing drugs. They're bringing crime. They're rapists. And some, I assume, are good people' (Associated Press, 2015: n.p.). Trump's views disturbingly articulate widely held ideas among a segment of the US general public that casts migrants as criminals and equates migration with illegality. Where have these associations come from? How are criminality and illegality mobilised in US immigration enforcement? How do they link immigration to larger carceral regimes? How are criminality and carcerality materialised in detention practices? What are the broader implications for understanding and researching the mobility of carceralisation in society as a whole? In this chapter we consider these questions drawing insight from scholarship on carceral geography (Loyd et al., 2012; Moran, 2015; Moran et al., 2013; Peck, 2003; Philo, 2012), which investigates issues of human security with a particular focus on spaces used to detain and incapacitate supposedly problematic groups. We also draw upon work by 'new mobilities' scholars including Cresswell (1999, 2010, 2012) and Sheller and Urry (2006). Finally, we employ research and reports on migrant detention in the US (ACLU, 2012; Doty and Wheatley, 2013; DWN, 2015; Freeman and Major, 2012; HRW, 2009; Meissner et al., 2013) as well as our ongoing research on the internal micro-economies of migrant detention (Conlon and Hiemstra, 2014; Hiemstra and Conlon, 2016). We argue that criminality in the context of immigration law and immigration detention is mobilised as an iteration of the carceralisation of society (Peck, 2003). Not only does this process shape problematic conceptions of migrants and mobility across borders like those articulated by Donald Trump and his retinue. In addition, echoing McCann (2010), who identifies a need to examine the 'mobility of ideas' (cited in Cresswell, 2012: 651), we propose that the movement of carceral ideology and practices from spaces of confinement, such as prisons and immigrant detention, into wider public spheres has profound implications for understanding the 'politics of mobility' (Cresswell, 2010) and the production of exclusion for other social groups in contemporary society. We also demonstrate how tracing the evolution of hegemonic ideas can deepen understanding in carceral mobilities scholarship.

The chapter develops as follows: first we provide an overview of developments in US immigration law in recent decades to identify how criminality, illegality, and immigration have been conjoined in a legal landscape that shapes official responses to and popular perceptions of migrants in North America. Following this we examine how criminality and carcerality are mobilised and materialised in the immigration detention system. We highlight key points where prison and detention systems intersect to show how they are instrumental in shaping migrant experiences and producing popular misconceptions where distinctions between criminality and migrant identity collapse. We then examine some of the broader effects and implications of carceralisation and the mobilisation and impact of carceral ideology and practices in wider spheres, registering how this system impacts family and support networks, for example. In the conclusion we reflect more broadly on impacts of the mobility of the carceral system for other social groups, for counter-mobilisations, and for scholarship.

Mobilising the carceral system: Criminalising mobility and migrants

The linking of immigrants with illegality and criminality has a long history in the US, as elsewhere (Cresswell, 2006). Mobile bodies are at once suspicious and outsiders, transgressive simply by defying norms of fixity and rootedness in place. For instance, early research in US sociology highlighted distinctive representations of mobility and class that overlay 'duty' versus 'delinquency' with representations of the 'globetrotter' and the 'hobo' respectively (Burgess, cited in Cresswell, 2010: 22). Indeed, Coleman (2012) argues that from the beginning of US immigration law, initiated with the Chinese Exclusion Act in 1882, immigrants have been viewed with suspicion. This has been instrumental to 'the production of immigrant il-legality, or extra-legality' (Coleman, 2012: 420) so that non-citizens are both bound by and excluded from the US legal system. As Ngai (2004) puts it, this system produces undocumented immigrants as 'impossible subjects' who are marginalised and excluded within society precisely because of their mobility. Here, we focus on recent history and immigration laws – especially since the late 1980s – when the implementation of a sweeping array of laws and programmes helped construct and codify immigrants as criminals. Whereas immigration detention in the US is technically administrative and not punitive, in other words it falls within civil and not penal law, this distinction paradoxically affords fewer legal protections to those detained for immigration violations than for criminal acts (Golash-Boza, 2010; Hernández, 2008). As Meissner et al. observe: 'noncitizens today … encounter the criminal justice system in unprecedented numbers and situations' (2013: 92).

In 1988, the *Anti-Drug Abuse Act*[1] was introduced into law and this piece of legislation brought aggravated felonies, a term and category of crime unique to immigration law and carrying a penalty of deportation if convicted, into existence.

Initially, four felony crimes were designated as aggravated felonies.[2] A subsequent round of punitive immigration laws in the 1990s expanded the number and type of crimes in this category.[3] By 2010, over 50 crimes were so designated (Immigration Policy Center, 2012; Miller, 2003; Zuniga, 2010). This has led some immigration judges to note that numerous 'non-violent, fairly trivial misdemeanors are considered aggravated felonies under our immigration laws' (Marks and Slavin, 2012: 91). Once their sentence is served in the criminal justice system, migrants are transferred to immigration detention, which is mandatory pending deportation. An offence can also be categorised as an aggravated felony retroactively, meaning that non-citizens (including legal permanent residents) who may have committed minor crimes years prior forever run the risk of deportation. In short, aggravated felony laws mete out severe penalties to immigrants while also codifying them in the public imaginary as criminals.

In addition, changes to the legal landscape that girds immigration have resulted in a massive increase in the number of immigrants – including those with legal status – coming into contact with the US criminal justice and prison systems, in ways that further mobilise ideologies that construct migrants as criminals. These changes, described below, include increasingly harsh punishments for unlawful entry and re-entry to the US, sentencing immigrants en masse for these 'offences', and sharing of information about an immigrant's status between criminal justice and immigration enforcement divisions of government.

While statutes governing unauthorised entry to the US have been in place for many years, penalties associated with mobility, for 'illegal entry' and 'illegal re-entry' specifically, have become increasingly harsh in recent history. When first introduced in the early twentieth century, illegal entry – where an immigrant is apprehended while attempting to enter the US at an unauthorised port of entry – carried a misdemeanour charge with a jail sentence of up to one year. Illegal re-entry – when an immigrant is apprehended after being deported – carried a felony charge with a prison sentence of up to two years.[4] In 1952 and again in 1988 punishment for these crimes expanded in space and time. Initially, borders and border crossing points, in other words points of entry into the US, were the only sites where a non-citizen could be apprehended and subsequently convicted on a charge of 'illegal re-entry'. Since 1952, however, unauthorised *presence* in the country has been the basis for a criminal conviction under illegal re-entry laws (Keller, 2012). The prison sentence for 'illegal re-entry' has also increased and today carries a maximum sentence of 20 years (US Sentencing Commission, 2015). More recently, in 2005, a policy initiative called *Operation Streamline* was implemented resulting in the criminalisation of migrants en masse for unauthorised entry or re-entry along certain sectors of the US–Mexico border. Prior to *Streamline* a first time apprehension at the border resulted in return or removal from the US without a criminal conviction while only migrants who had re-entered after deportation or with a criminal record were referred for prosecution. *Operation Streamline* stipulates that migrants who are apprehended at designated border crossing sites, primarily in Arizona and Texas, are 'arrested, detained while awaiting trial, prosecuted with a misdemeanour or felony charge, incarcerated in the federal justice system, and finally deported' (Buentello et al., 2010: 3). A notable characteristic of the programme is its use of 'rapid-fire group trials' (HRW, 2013: 35) where upwards of 70 migrants are prosecuted at a given time (Burridge, 2011; Lowen, 2016; Meissner et al., 2013). Also noteworthy is that while *Streamline* is supposed to include provisions for asylum seekers or humanitarian relief, reports indicate that migrants within these categories are regularly referred for prosecution (HRW, 2013; Office of Inspector General, 2015). In effect, regardless of conditions that might necessitate mobility in the first place, increasing numbers of migrants are collapsed into a single category that is classed as 'criminal'.

Criminal prosecutions for violations of immigration laws have now skyrocketed. Over 90,000 people were prosecuted for unauthorised entry or re-entry in 2013 compared with 12,000 prosecutions in 2002 (Grassroots Leadership, 2014).[5] Today, Customs and Border Patrol (CBP), the unit within the Department of Homeland Security that implements Streamline, along with ICE together refer more cases for criminal prosecution than do all Department of Justice law enforcement agencies combined (Meissner et al., 2013: 94–95). Moreover, the programme has swept up into the carceral system huge numbers of migrants with no criminal history whatsoever and also exacts consequences that have profound and far-reaching impacts (Lowen, 2016). For instance, reports indicate that significant numbers of migrants crossing the US–Mexico border do so in order to reunite with family members already in the US (HRW, 2013). With a criminal conviction and subsequent deportation an individual is barred from ever returning to the US legally. This separates families and loved ones, increases economic vulnerability, and ensures that any attempt to return to the US requires the use of unauthorised means and routes, which further exacerbates vulnerability while also enriching smugglers' and traffickers' coffers. Put differently, a system that is intended to curtail movement by linking it with criminality has effectively mobilised and invigorated illicit practices. With this we see how carceral ideology 'moves' in different spheres and begets practices that more or less guarantee migrant incarceration.

Other nationwide programmes have contributed to the mobility of the border by shifting the scale of as well as the agents involved in immigration enforcement, resulting in an even wider net of individuals coming into contact with ICE, categorised as criminal alien and ultimately expelled from the US. Under Section 287(g) of the *Immigration and Nationality Act*,[6] local law enforcement agencies could be authorised by federal authorities to enforce federal immigration law, which effectively eroded long-standing models of immigration enforcement as solely within the purview of federal authorities (Miller 2003; Coleman and Kocher 2011). The perversely named 'Secure Communities' programme, in place from 2008 to 2015, institutionalised information sharing related to an individual's criminal history and immigration status between all levels of law enforcement agencies. Arrested individuals who were found to be in the US without authorisation were transferred to ICE custody, subsequently referred for criminal prosecution, and eventually placed in deportation proceedings under ICE's jurisdiction. Secure

Communities resulted in massive expansion in the number of individuals screened for immigration status and subsequently deported (Meissner et al., 2013: 110). Under mounting criticism and increasing numbers of cities and states refusing to participate citing damage to relations between immigrant communities and local police, in 2015 the Obama administration replaced Secure Communities with the 'Priority Enforcement Program', or PEP (Feliz, 2015). PEP is supposed to emphasise the use of detainers and transfer to ICE custody only for high-level 'priority' offenses. While it is too early to assess if PEP actually works any differently than Secure Communities, critics remain highly sceptical, and significant numbers of law enforcement agencies are still defying orders to participate (Feliz, 2015). Regardless, the conflation of state and local level criminal justice and immigration enforcement has been firmly established.

It is clear that the number of migrants classified as criminals, coming into contact with the criminal justice system, or imprisoned has ballooned in recent decades. Hand in hand with these developments is the massive expansion and privatisation of the immigration detention system, which, as we discuss in more detail further on, often shares space and services with the prison system. A growing body of critical work highlights the influence of private corporations in political lobbying efforts to expand what is now referred to as 'crimmigration' (Stumpf, 2006) and to ensure the continued growth of immigration detention (Cervantes-Gautschi, 2010; Conlon and Hiemstra, 2014; Doty and Wheatley, 2013; Golash-Boza, 2009; Miller, 2003; Mitchelson, 2014).[7] The detention system is comprised of approximately 250 facilities in which over 400,000 non-citizens are detained annually. Detention facilities can be owned and operated by federal, state or county governments, by a private company, or by some combination of these entities (Doty and Wheatley, 2013; Meissner and Kerwin, 2009). Two points are important for our purposes here: first, the range of facilities used for immigration detention includes current and former jails and prisons as well as privately owned and operated facilities, and second, regardless of who runs a facility, privatisation permeates the detention system as even where facilities are government owned major components of operation are often contracted out to the private sector. This gives some indication of various stakeholders – politicians, legal entities, law enforcement, local governments, and private corporations – that have a vested interest in shaping public opinion that casts migrants as criminals in an effort to garner support for further expansion of detention (Doty and Wheatley, 2013; Golash-Boza, 2009; Miller, 2003). In short, a confluence of politics, law, and privatisation has mobilised ideologies that entwine migrants with criminality and place them firmly within the grip of the carceral system. How is this system materialised on a day-to-day basis and in the experiences of migrants who are ensnared within immigration detention? We turn to this question in the next section.

Materialising criminality within immigration detention

Given the ways migrant status and criminality overlap within an ideological and legal landscape, it is hardly surprising that sites, spaces, and services for prison populations and migrant detainees overlap. This is evidenced in the expanding proportion of the prison population serving time for immigration related offenses, a trend that reached a 'tipping point [in 2009] when more people entered federal prison for immigration offenses than for violent, weapons, and property offenses combined' (ACLU, 2014: 2) and continues to rise today. It is also evidenced in the use of jail space and former prisons as detention facilities for those awaiting an immigration hearing or deportation (ICE, 2011; Hiemstra and Conlon, 2016). Indeed, some commentators point out that immigration detainees represent an important revenue stream for state and local governments where they serve as a 'replacement' for declining state prison populations (Ackerman and Furman, 2013; Justice Policy Institute, 2011).

Carcerality is also materialised in spaces of detention and for migrants throughout the system in almost every facet of daily life. As Schriro notes, facilities 'vary in age and architecture. Quite a few do not have windows. A number consist of single and double-celled units' (2009: 21) and many are in remote areas; this limits access to transportation for family, advocates, and other potential visitors. In state and county jails where migrants and criminal populations are held in the same space, they are usually housed separately from each other; however, services are provided to the facility as a whole or subcontracted to private corporations that do not differentiate between the institutionalised populations they serve. For instance, food service provision contracts treat migrant and corrections populations the same way and exercise heightened security practices in their operations regardless of whether in a prison or detention facility. Similarly, commissary systems operate according to a corrections institution logic; with this the types of goods available for purchase and the practice of accessing the commissary are the same (for discussion of these 'micro-economies' within detention see Conlon and Hiemstra, 2014; Hiemstra and Conlon, 2016). Thus, with few exceptions, detention centres operate just like prisons, according to logics of security, surveillance, command, and control. Migrant encounters with this carceral system are even further reinforced in the discourse and day-to-day culture of detention. Detainee handbooks offer significant insight here.[8]

Detainees are provided with a detainee handbook as part of their "orientation" to a facility, which lays out rights and responsibilities pertaining to conditions and day-to-day life in the facility. Provision of handbooks accords with guidelines set out by the Performance Based National Detention Standards (PBNDS), which were developed in 2008 and revised in 2011 in response to a review and recommendations on improving detention standards (see Schriro, 2009). Importantly, detention standards, including guidelines for preparation of detainee handbooks draw, in part, from standards identified by the American Correctional Association; as such they carry 'criminal incarceration policies and practices into the arena of immigration detention' (Schriro, 2009: 16). Facilities routinely fail to comply with 'requirements' laid out in the PBNDS and, indeed, there are currently no penalties for non-compliance.[9] Detainee handbooks distributed by facilities we have researched follow a template provided in a section of the PBNDS (ICE, 2011: Section 6: 388–391) and in this respect can be seen to nominally comply with the standards. In this section, we highlight ways in which elements of the handbook – specifically admission procedures and aspects of general orientation, the classification system, and the disciplinary process – serve to represent and materialise migrants as a potential threat or criminal. In so doing, they rationalise and reinforce expressions and experiences of carcerality.

As detailed in handbooks, upon admission to a facility, an individual must swap their 'civilian' clothes for a prison jumpsuit and relinquish all personal items and money in their possession with the exception of a wedding band, which is permitted. These items are stored within the facility until the detainee is transferred, released, or more often, deported. Detainees also undergo a medical exam and an assessment to determine a 'classification level'. In this process detained migrants are effectively stripped of their identity in much the same way as a prisoner entering the corrections system. Expressions of individuality or affiliation are vetoed, expectations about behaviour and demeanour – such as how detention centre staff should be addressed – are set out as rules, and consequences for non-compliance are detailed at length. With this, an understanding of migrants as different, as potential threat, and in essence, as criminal is spatialised and becomes manifest immediately.

The detainee classification system is used to establish 'security needs' (Community Education Centers (CEC), no date: 8),[10] which is to say, level of control, for individuals upon admission to a facility.[11] The classification system is quite elaborate with different levels (from low (1) to high (3)), different statuses (major and minor), as well as distinct categories for males and females, special needs (such as medical, mental health, or substance abuse), and juveniles. Where space within facilities allows it, differently classified groups are housed separately and commingling between certain groups is not permitted (e.g., level 1 and 3 detainees cannot mix). According to the detainee handbook, classification is assessed 'based on criminal behavior, criminal convictions, immigration history, disciplinary record, [and] current custody level' (Essex County, 2013: 8). For individuals detained at Delaney Hall, a facility that detains immigrants classified as 'low level' security risk who 'do not have a significant history of criminal behavior' (CEC, no date: 10) the initial classification assessment includes an eleven item list including interview questions to determine 'prior criminal and history of escape, aggressive or passive tendencies, gang membership, and criminal sophistication' (CEC, no date: 9).

We have described the classification system in detail because it illustrates that detainee criminality is assumed at admission and the classification system defines detained immigrants in accordance with the apparent degree of threat they represent. In other words, day-to-day life within detention space mobilises and reinforces migrant carcerality. Even where facilities are explicitly designated for individuals with no criminal record, the basis for initial assessment and classification presumes criminality. With this, it is, perhaps, not surprising that in the general public's view distinctions between migrant and criminal collapse together. It is also noteworthy that an individual's classification level is subject to revision at any time 'based on direct observation and supervision of ICE detainees' (Essex County,

2013: 8) and any review or reclassification could potentially result in a detainee's 'housing assignment being changed [… or potentially being placed] into a segregation unit' (Essex County, 2013: 9). As such, the classification system serves as a disciplinary tool that constructs detained migrants as unruly, which rationalises the use of detention as well as the degree of control exercised over detainees.

On the matter of discipline, a substantial proportion of the detainee handbook is given over to identifying actions that beget disciplinary action and to detailing the disciplinary process. Details related to discipline comprise almost twenty per cent of the 48-page handbook for the Essex County detention centre and 16 per cent of the Delaney Hall handbook where migrants are classified as 'low level' detainees. This is in contrast to 8 per cent of the information presented devoted to communication and family contact and less than 1 per cent to the law library and recreation services respectively, each of which is a required programme service according to the PBNDS.

The array of acts that constitute disciplinary violations is dizzying, including a 47-item list under 'prohibited acts' (Essex County, 2013: 4–6) an additional list of over 40 'minor violations' (2013: 15) and an even more extensive list of 'major violations' (2013: 17–18). Among 'prohibited acts' are: leaning on a stairs or near pillars, storage of commissary bought food for more than one week, and keeping garbage or plastic bags in cells. Behaviours that warrant disciplinary action as 'minor violations' include being insanitary or untidy or refusing to work, while 'major violations' include lying to a member of staff, fighting with another person, and possessing money or currency in excess of $50.00.

Consequences for violations range from loss of privileges, such as recreation or TV access for up to five days and confinement to cell for up to four hours for minor violations, to loss of privileges for up to 30 days, up to 15 days' disciplinary detention, which may include 'disciplinary segregation' (commonly known as solitary confinement in criminal justice settings), or, if necessary, filing criminal charges against the detainee. A number of reports indicate that despite elaborate details, the disciplinary process is uneven, unreasonably harsh, and unfair punishments are meted out regularly (see AFSC, 2015; Gonzalez, 2015). That discipline saturates detention as outlined here is remarkable for what is supposed to be a matter of civil law and an administrative process. As Schriro reports, the standards used in detention 'impose more restrictions and carry more costs than are necessary to effectively manage the majority of the detained population' (2009: 3–4), elaborating further, 'the majority of the population is motivated by the desire for repatriation or relief, and exercise remarkable restraint' (2009: 21). The great emphasis on discipline and the disciplinary process speak to the mobilisation and superimposition of command and control logic, common to corrections settings, in the shared space of immigrant detention. Moreover, this logic imagines and produces detainees as troublesome or potentially so, always on the verge of committing some disruptive act, which justifies treating them as though they are individuals incarcerated for criminal acts and similarly confined within a carceral space.

The mobility of a carceral logic and its materialisation within detention means there is little substantive distinction between detained migrants and an incarcerated criminal population. This also obscures distinctions between Latino/a immigrants more broadly and their conception as criminals (see Hernández, 2008). The mobilisation and materialisation of carcerality vis-à-vis immigration law and migrant detention, therefore, help to account for the question posed in this chapter's introduction, namely, from where have associations that cast migrants as criminals and equate migration with criminality and illegality come? An equally important question is where are these associations going? In other words, what are the broader effects and implications of migrant carcerality for contemporary society? And how might we, as scholars, researchers, and society respond? We reflect on these questions in the final section.

Concluding thoughts on the impact and implications of carceralisation

The mobility and materiality of carcerality in immigration law and detention exact a profound toll on migrants and their extended networks. Numerous reports attest to abysmal conditions and the debilitating impact of detention on the mental and physical health and well being of detainees (ACLU, 2012; Bosworth, 2014; Conlon and Hiemstra, 2014; Gill, 2016; HRW, 2009; Meissner et al., 2013). Scholars also show how 'ontological insecurity' (Katz, 2007), which is marked by a profound and perpetual 'state of anxiety about the future' (Katz, 2008: 6), takes root in immigrant communities in association with the threat of being picked up, detained, or deported either by law or immigration enforcement agents (Coleman and Kocher, 2011; Conlon, 2015; Harrison and Lloyd, 2012; Hiemstra, 2008). Anxieties extend to family and support networks, too. For instance, like prisons, detention facilities are often located in remote areas; also, detainees experience frequent transfers, routinely over hundreds of miles (Hiemstra, 2013; HRW, 2009). The burden associated with maintaining contact falls on family members outside detention and doing so necessitates hours of travel time as well as arranging time off from work or coordinating child care, as well as myriad other logistics. When a detainee is transferred there can be a gap in knowing where s/he is, and the practice immediately evokes concern about whether the individual has been moved or deported (Hiemstra, 2012). With this, we see that much of the 'weight' of migrant encounters with carcerality is borne by family and networks, at minimum, through perpetual worry and over-extensions of time and expense in efforts to maintain contact.[12]

This chapter has also offered a particular way of doing carceral mobilities research: through tracking and following the development and materialisation of dominant ideologies. We traced the manufactured nexus between migrants and criminality that shapes public opinion, bolsters popular support for pernicious anti-immigrant politics such as those seen in the 2016 US election campaign, and builds support for nativist and anti–human rights policies. Discourses of migrant criminality also deepen hostility toward migrants, which exacerbates the fears and insecurities that define everyday life for migrants. These effects are disturbing by themselves. They are also particularly troubling when we think about the implications of migrant carcerality vis-à-vis other social groups and in society broadly.

Writing over a decade ago, Peck observed, 'the prison system can be understood as one of the epicentral institutions of these neoliberalized times' (2003: 226). In his review of scholarship that critically examines the multi-faceted dynamics of neoliberal policies, Peck identified a need to eschew linear or pre-conceived ideas about the 'shrinking state' or wholesale advance of free markets under neoliber-alism, and called for attention to 'the social/penal frontier as an active zone of statebuilding' (2003: 230) in the current era. He goes on to outline scholarship that identifies growing swaths of social groups, including but not limited to youth, working-class men, minorities, the homeless, the disabled, as well as immigrants, who are now subjected to 'new forms of micro-social intervention' (2003: 227) and regulation that amounts to carceral control. In this vein, our examination of immigration enforcement in the US highlights how carceral ideology and prac-tice have been mobilised beyond prison to immigration detention and beyond this again to extend the social/penal frontier into wider spheres. We have shown, for instance, that the extension of aggravated felony laws retroactively means not only migrants but also legal permanent residents can now be swept into the carceral system. We have also shown how screening programmes and data sharing proto-cols in the name of immigration enforcement mean massively expanded numbers of individuals – migrant and non-migrant alike – are subject to monitoring and potential control. These extensions and mobilisations of the social/penal frontier also extend what Cresswell (2010) describes as the 'politics of mobility' to show that power relations embedded within embodied experiences, migrant identity, and mobility also operate at scales of ideology and institutional practice.

Yet another dimension of the broader mobility of carceralisation can be seen in its dynamic morphing and expansion in other spheres. Peck's analysis suggests that carceralisation is simultaneously a new form of statecraft and a growth indus-try. This is certainly the case with immigration enforcement and can be seen in yet another impact of detention for migrant families. As more and more migrants are detained or deported, the number of children in the child welfare system has expanded rapidly. Research completed by the Applied Research Center (ARC) estimated at least 5,100 children of detained or deported migrants in the foster care system in 2011. The same report noted this figure was expected to increase over five years, with '15,000 *more* children [facing] threats to reunification with the detained and deported mothers and fathers' (2011: 4, italics in original). When it comes to deportation specifically, one report observes that of migrants who were deported in the first six months of 2011, twenty-five per cent were the parent of a US citizen (Wessler, 2011). In these situations there is 'a systematic bias against reunifying children with parents in other countries' (ARC, 2011: 6). Beyond the obvious emotional and social toll, this practice exposes a new group, US citizen children, to institutionalisation and some suggest also sets them on a path for future encounters with the carceral system as 'children who wind up in the child welfare system have a higher incidence of involvement with the criminal justice system than the general youth population' (Ackerman and Furman, 2013: 259). Just as the carceral system ensnares the children of detained and deported adults within its grip, it seems logical to conclude that its weed-like potential is imminent for other social groups. McCann (2010) identifies a need to scrutinise how ideas shift as they move. Our analysis in this chapter represents one step in this process by identifying how carceral ideology has been mobilised and, in turn, moves from one sphere to another. More work is needed, however, to carefully track, trace, and 'reveal how things change in transit', as Cresswell (2012: 651) observes.

This leads to a further, final reflection on how we in society and as scholars can respond. In other words, as punitive control and exclusion of ever-increasing masses of society continue, what opportunities for counter-mobilisation against carceralisation are possible? In formulating a tentative response to this question, we return to Donald Trump's presidential candidacy, which exemplifies an especially bombastic and putrid manifestation of carceral thinking. At the time of writing, a number of media reports indicated significant numbers of naturalised immigrants in the US are registering to vote with the explicit aim of voting against Trump and what he stands for (Schlesinger, 2016), as well as immigrants seeking citizenship just to vote (Preston, 2016). Additionally, at a political rally in Chicago, Illinois, which was shut down for fomenting unease and potential violence toward attendees who oppose the rhetoric of Trump's campaign, a number of accounts identify the beginnings of a coalition of disparate groups, among them minorities, women, and young people, who reject being maligned, marginalised, and subjected to unjust control (Ollstein and Lee, 2016). Such movements, nascent and fragile as they may be, suggest the emergence of a much-needed counter-mobilisation that opposes carceralisation writ large (see Loyd et al., 2012). If, as scholars, we wish to align our work with these counter-mobilisation efforts, we must recognise, as Cresswell (1999: 177) does, that mobility is 'the most basic form of intentionality' and work in mobility studies must continuously work toward exposing and extending analyses of the 'politics of mobility' in this carceral age.

Notes

1 Anti-Drug Abuse Act of 1988, Pub. L. 100-690, 102 Stat. 4181, 100th Cong., 2nd sess., (18 Nov 1988), Sec. 7342.
2 These included murder, rape or sexual abuse of a minor, money laundering, and crimes of violence (see Wiegand, 2011).
3 *Omnibus Consolidated Appropriations Act*, 1997, Pub. L. 104-208, 110 Stat. 3009-664 (30 September 1996) (incorporating the *Illegal Immigration Reform and Immigrant Responsibility Act* (IIRIRA) and *Anti-terrorism and Effective Death Penalty Act* (AEDPA) Pub. L. no. 104-32, 110 Stat. 1214, 104th Cong, 2nd session (24 April 1996).
4 Illegal entry and re-entry were introduced with passage of immigration laws in 1929 (4 March 1929). Earlier laws included the 1918 Passport Act (entry law) and 1798 Aliens Act (re-entry laws). For detailed discussion of these and the current status of illegal entry and re-entry laws, see Keller (2012).
5 Thanks to Matthew Lowen for alerting us to this report.

6 Although adopted in 1996, 287(g) was not widely implemented until after 11 September 2001.

7 CCA, Geo Group, and MTC are the 'top three' corporations involved in immigration detention. Companies with a smaller share of the privatized immigration detention industry include Community Education Centers, Emerald Connections and LCS Corrections (DWN, 2011).

8 This section draws on research from a broader project examining the internal micro-economies of immigration detention in the greater New York City metropolitan area. For that project, we have gathered data from three sources: a series of Freedom of Information (FOIA) requests submitted at the federal level and comparable requests (Office of Public Record Act, OPRA) at the state level, 15 semi-structured interviews with individuals who have inside knowledge and experience of detention facilities in some capacity (e.g. lawyers, activists, visitors, and a former detainee), and a review of published reports on conditions within detention. Here we focus on detainee handbooks for two facilities: Essex County Correctional Facility is a county jail that also has a contract with ICE to detain migrants, and Delaney Hall is owned by Essex County and operated by a private contractor, Community Education Centers. Further detail can be found in Conlon and Hiemstra (2014) and Hiemstra and Conlon (2016).

9 Inspections reflect a 'checklist culture' (NIJC, 2015: n.p.) that records the presence or absence of policies but not how they are implemented. Facility inspections based on these standards have been deemed a 'sham' with 'inspectors – employed by ICE directly or via subcontracts – [who] engage in pre-planned, perfunctory reviews of detention facilities that are designed to result in passing ratings and to ensure local counties and private prison corporations continue to receive government funds' (NIJC, 2015: n.p.).

10 Detainee handbooks for Delaney Hall and Essex County Correction Facility are available from the authors.

11 Thanks to Ella Graham and Emily Cowling at the University of Leeds for research assistance on detainee classification.

12 While not the focus of this chapter it is important to note that the developments in law and discourse discussed here also have enormous monetary implications. Costs to government and taxpayers in terms of fiscal resources and labour time as well as record level profits generated in the immigration industrial complex are detailed in a growing number of reports, see for example: AFSC (2015); Buentello et al. (2010); Conlon and Hiemstra (2014).

References

Ackerman A and Furman R (2013) The criminalization of immigration and the privatization of immigration detention: implications for justice. *Contemporary Justice Review* 16(2): 251–261.

American Civil Liberties Union) (2012) *Prisoners of Profit: Immigrants and Detention in Georgia.* Available at: www.acluga.org/files/2713/3788/2900/Prisoners_of_Profit.pdf

ACLU (American Civil Liberties Union) (2014) *Warehoused and Forgotten: Immigrants Trapped in our Shadow Private Prison System.* Available at: www.aclu.org/sites/default/files/assets/060614-aclu-car-reportonline.pdf

AFSC (American Friends Service Committee) (2015) *23 Hours in a Box: Solitary Confinement in New Jersey Immigration Detention.* Available at: https://afsc.org/sites/afsc.civicactions.net/files/documents/23%20Hours%20in%20the%20Box.pdf

ARC (Applied Research Center) (2011) *Shattered Families: The Perilous Intersection of Immigration Enforcement and the Child Welfare System.* Available at: www.raceforward.org

Associated Press (2015) What did Trump say about immigrants? *Boston Globe*, 29 June. Available at: www.bostonglobe.com

Bosworth M (2014) *Inside Immigration Detention*. New York, NY: Oxford University Press.

Buentello T, Carswell S, Hudson N and Libal B (2010) *Operation Streamline: Drowning Justice and Draining Dollars Along the Rio Grande*. Available at: www.grassrootsleadership.org

Burridge A (2011) Differential criminalization under Operation Streamline: Challenges to freedom of movement and humanitarian aid provision in the Mexico-U.S. borderlands. *Refuge* 26(2): 78–91.

Cervantes-Gautschi P (2010) Wall Street and the criminalization of immigrants. *Americas Program*, 10 June. Available at: www.cipamericas.org/archives/3304

Coleman M (2012) Immigrant il-legality: Geopolitical and legal borders in the U.S., 1882–present. *Geopolitics* 17(2): 402–422.

Coleman M and Kocher A (2011) Detention, deportation, devolution and immigrant incapacitation in the US, post 9/11. *The Geographical Journal* 177(3): 228–237.

Community Education Centers (CEC) (no date) *Delaney Hall Detainee Handbook*. Essex County, NJ. [on file with authors].

Conlon D (2015) Ontological insecurity, migrant journeys and the migration industry. Paper presented at the Annual Meeting of the Association of American Geographers, Chicago, IL.

Conlon D and Hiemstra N (2014) Examining the everyday micro-economies of immigrant detention in the United States. *Geographica Helvetica* 69(5): 335–344.

Cresswell T (1999) Embodiment, power and the politics of mobility: The case of female tramps and hobos. *Transactions of the Institute of British Geographers* 24(2): 175–192.

Cresswell T (2006) *On the Move: Mobility in the Modern Western World*. Abingdon: Routledge.

Cresswell T (2010) Towards a politics of mobility. *Environment and Planning D: Society and Space* 28(1): 17–31.

Cresswell T (2012) Mobilities II: Still. *Progress in Human Geography* 36(5): 645–653.

Doty RL and Wheatley ES (2013) Private detention and the immigration industrial complex. *International Political Sociology* 7(4): 426–443.

DWN (Detention Watch Network) (2011) *The Influence of the Private Prison Industry in the Immigration Detention Business*. Available at: www.detentionwatchnetwork.org/sites/default/files/reports/DWN%20Private%20Prison%20Influence%20Report.pdf

DWN (Detention Watch Network) (2015) *Banking on Detention: Local Lock-up Quotas and the Immigrant Dragnet*. Available at: www.detentionwatchnetwork.org/sites/default/files/Banking_on_Detention_DWN.pdf

Essex County (2013) *Essex County Correctional Facility ICE Detainee Handbook*. Essex County, NJ: Department of Corrections. [on file with authors].

Feliz W (2015) DHS faces challenges as it rolls out the priority enforcement program. *Immigration Impact*, 4 August. Available at: http://immigrationimpact.com/2015/08/05/priority-enforcement-program-launch/

Freeman S and Major L (2012) Immigration incarceration: The expansion and failed reform of immigration detention in Essex County, NJ. *NYU School of Law Immigrant Rights Clinic*, March. Available at: www.afsc.org/sites/afsc.civicactions.net/files/documents/ImmigrationIncarceration2012.pdf

Gill N (2016) *Nothing Personal? Geographies of Governing and Activism in the British asylum System*. Oxford: Wiley-Blackwell.

Golash-Boza T (2009) The immigration industrial complex: Why we enforce immigration policies destined to fail. *Sociology Compass* 3(2): 295–309.

Golash-Boza T (2010) The criminalization of undocumented migrants: Legalities and realities. *Societies Without Borders* 5(1): 81–90.

Gonzalez S (2015) Alone and isolated, the punishment piles on for immigration detainees. *WNYC News*. Available at: www.wnyc.org/story/solitary-confinement-immigrant-detention/

Grassroots Leadership (2014) *Shadow Report of Grassroots Leadership and Justice Strategies to The International Convention on the Elimination of All Forms of Racial Discrimination Regarding Criminal Prosecutions of Migrants for Immigration Offenses and Substandard Privately Operated Segregated Prisons.* Available at: http:// grassrootsleadership.org/

Harrison J L and Lloyd SE (2012) Illegality at work: Deportability and the productive new era of immigration enforcement. *Antipode* 44(2): 365–385.

Hernández DM (2008) Pursuant to deportation: Latinos and immigrant detention. *Latino Studies* 6(1–2): 35–63.

Hiemstra N (2008) Spatial disjunctures and division in the New West: Latino immigration to Leadville, Colorado. In: Jones, RC (ed) *Immigrants Outside Megalopolis: Ethnic Transformation in the Heartland.* Lanham, MD: Lexington Books, 89–113.

Hiemstra N (2012) Geopolitical reverberations of US migrant detention and deportation: The view from Ecuador. *Geopolitics* 17(2): 293–311.

Hiemstra N (2013) 'You don't even know where you are': Chaotic geographies of US migrant detention and deportation. In: Moran D, Gill N and Conlon D (eds) *Carceral Spaces: Mobility and Agency in Imprisonment and Migrant Detention.* Farnham: Ashgate, 57–75.

Hiemstra N and Conlon D (2016) Captive consumers and coerced labourers: Intimate economies and the expanding US detention regime. In: Conlon D and Hiemstra N (eds) *Intimate Economies of Immigration Detention: Critical Perspectives.* Abingdon: Routledge, 123–139.

HRW (Human Rights Watch) (2009) *Locked Up Far Away: The Transfer of Immigrants to Remote Detention Centers in the United States.* New York, NY: Human Rights Watch.

HRW (Human Rights Watch) (2013) *Turning Migrants into Criminals: The Harmful Impact of US Border Prosecutions.* New York, NY: Human Rights Watch.

ICE (2011) Operations manual ICE performance-based national detention standards. *Department of Homeland Security.* Available at: www.ice.gov/detention-standards/2011

Immigration Policy Center (2012) *Aggravated Felonies: An Overview.* Washington, DC: Immigration Policy Center.

Justice Policy Institute (2011) *Gaming the System: How the Political Strategies of Private Prison Companies Promote Ineffective Policies.* Washington, DC: Justice Policy Institute.

Katz C (2008) Childhood as spectacle: Relays of anxiety and the reconfiguration of the child. *Cultural Geographies* 15(1): 5–17.

Katz C (2007) Banal terrorism: Spatial fetishism and everyday insecurity. In Gregory D and Pred A (eds) *Violent Geographies: Fear, Terror, and Political Violence.* Abingdon: Routledge: 349–361.

Keller D (2012) Rethinking illegal entry and re-entry. *Chicago Law Journal* 44: 65–139.

Marks LD and Slavin DN (2012) A view through the looking glass: How crimes appear from the immigration court perspective. *Fordham Urban Law Journal* 39(1): 91–119.

Lowen M (2016) Intimate encounters with immigrant criminalisation in Arizona. In: Conlon D and Hiemstra N (eds) *Intimate Economies of Immigration Detention: Critical Perspectives.* Abingdon: Routledge, 187–202.

Loyd J, Mitchelson M, and Burridge A (2012) *Beyond Walls and Cages: Prisons, Borders and Global Crisis.* Athens, GA: University of Georgia Press.

McCann E (2010) Urban policy mobilities and global circuits of knowledge: Toward a research agenda. *Annals of the Association of American Geographers* 101(1): 107–130.

Meissner D and Kerwin D (2009) DHS and immigration: Taking stock and correcting course. *Migration Policy Institute.* Available at: file:///C:/Users/Owner/Downloads/ DHS_Feb09%20(1).pdf

Meissner D, Kerwin D, Chisti M and Bergeron C (2013) *Immigration enforcement in the United States: The Rise of a Formidable Machinery.* Washington, DC: Migration Policy Institute.

Miller TA (2003) Citizenship & severity: Recent immigration reforms and the new penology. *Georgetown Immigration Law Journal* 17(4): 611–666.

Mitchelson ML (2014) The production of bedspace: Prison privatization and abstract space. *Geographica Helvetica* 69(5): 325–333.

Moran D (2015) *Carceral Geography: Spaces and Practices of Incarceration*. Farnham: Ashgate.

Moran D, Gill N and Conlon D (2013) (eds) *Carceral Spaces: Mobility and Agency in Imprisonment and Migrant Detention*. Farnham: Ashgate.

Ngai MM (2004) *Impossible Subjects: Illegal Aliens and the Making of Modern America*. Princeton, NJ: Princeton University Press.

NIJC (National Immigrant Justice Center) (2015) *Lives in Peril: How Ineffective Inspections Make ICE Complicit in Detention Center Abuse*. Chicago, IL: NIJC.

Office of the Inspector General, Department of Homeland Security (2015) *Streamline: Measuring its Effect on Illegal Border Crossing*. Washington, DC: Office of Public Affairs.

Ollstein A and Lee EY-h (2016) How activists mobilized to shut down Trump in Chicago. *ThinkProgress.org*, 12 March. Available at: http://thinkprogress.org/politics/2016/03/12/3759465/how-activists-mobilized-chicago/

Peck J (2003) Geography and public policy: Mapping the penal state. *Progress in Human Geography* 27(2): 222–232.

Philo C (2012) Security of geography/geography of security. *Transactions of the Institute of British Geographers* 37(1): 1–7.

Preston J (2016) More Latinos seek citizenship to vote against Trump. *The New York Times*, 7 March 7. Available at: www.nytimes.com/2016/03/08/us/trumps-rise-spurs-latino-immigrants-to-naturalize-to-vote-against-him.html?_r=0

Schlesinger R (2016) Hillary's secret weapon: Trump. *U.S. News and World Report*, 18 March. Available at: www.usnews.com/news/the-report/articles/2016-03-18/whites-arent-the-only-voters-donald-trump-will-turn-out

Sheller M and Urry J (2006) The new mobilities paradigm. *Environment and Planning A* 38(2): 207–226.

Schriro D (2009) Immigration detention overview and recommendations. *U.S. Immigration and Customs Enforcement, Department of Homeland Security*. Available at: www.ice.gov/doclib/about/offices/odpp/pdf/ice-detention-rpt.pdf

Stumpf J (2006) The crimmigration crisis: Immigrants, crime, and sovereign power. *American University Law Review* 56(2): 367–419.

US Sentencing Commission (2015) *Report on Illegal Reentry Offenses*. Washington, DC: USSC.

Wessler S (2011) US deports 46K parents with citizen kids in just six months. *Colorlines*. Available at: www.colorlines.com/articles/us-deports-46k-parents-citizen-kids-just-six-months

Wiegand C III (2011) *Fundamentals of Immigration Law*. Available at: http://permanent.access.gpo.gov/gpo30265/Fundamentals_of_Immigration_Law.pdf

Zuniga B (2010) *Aggravated Felony Case Summary*. San Antonio, Texas: Executive Office for Immigration Review (EOIR).

9 On 'floaters'

Constrained locomotion and complex micro-scale mobilities of objects in carceral environments

Anna Schliehe

Introduction

The mobility of objects and detainees' possessions in confined space is a largely overlooked dynamic in the prison context. Yet, objects and their movements tell tales of inner social worlds and complex interplays of care and control. Due to their high levels of mobility, many objects – ranging from mundane everyday supplies to prohibited contraband – signify the semipermeable nature of prison environments as they constantly cross boundaries of inside and outside. In an imaginary, as well as material form, they are markers of inner regimes but also evoke atmospheres and symbolic perceptions of closed space. Being interested in their 'loco-motional' characteristics means to analyse an object's change in position over time, as well considering the underlying power that influences this action and the expedients involved in accompanying spatial practices. Locomotion, underlines here 'locus' the place and displacement in a variety of movements.[1] The alteration of material and symbolic spaces through tactical use of objects has long been on the agenda of object-related theories and philosophy, and has been adopted by human geographers in a range of sub-disciplines.

While this chapter is conceptually positioned at the intersection of mobilities (Cresswell and Merriman, 2011; Gill, 2013; Urry, 2007) and object-oriented geographical enquiry (Bennett, 2004; Harman, 2005; Jackson, 2000; Latour, 2005; Shaw, 2012; Whatmore, 2006), it also introduces Goffman's (1991) work on the importance of objects in what he calls the 'underlife of the institution'. While being neither associated with mobilities nor object research, Goffman's micro-scale account of the use of objects in the formation of the prisoner's identity and sense of self proves too applicable to the empirical material to dismiss. His study of institutional underlife is based on analysing how prisoners and other inmates develop adaptive practices to integrate into the prison regime. In this context, objects play an overlooked but important role that has been touched on by Baer (2005) in his carceral geographic work on the tactical use of objects in creating personal space in prison environments. The empirical material reveals that it is not just objects that can be 'force-full' (Shaw and Meehan, 2013) but also imaginary or internal objects (also Sayers, 2000; Winnicott, 1986). In the social world of prisons, authorised as well as unauthorised objects hold multiple layers of meaning, and can thus be powerful entities creating difference to everyday life. Inherently connected to the locomotion of objects is their effect on prisoner roles and identities that are held both individually, and assigned institutionally.

The data presented here reflects views on the importance of objects drawing on in-depth qualitative interviews with young female detainees and members of staff. As part of a more extensive research project, objects were often talked about in passing, shown to the researcher while walking around the prison or pointed out by staff over a cup of coffee. These informal encounters with objects – often witnessing crucial movement and assignation of meaning by chance – opened up for analysis a micro-scale that reveals the changeable, subtle, and subversive nature of institutional life and underlife. From the small resistance of wearing one's own clothing at unauthorised times to losing all personal items in a status downgrade, or attempting to smuggle a handmade card to someone in segregation, the geographies of floating[2] objects are a constant but unnoticed companion to prison life.

In the following sections I provide a short overview of object mobility in geographical research, briefly touching on object-oriented ontology (OOO) and object relations theory (ORT). The main conceptual frame will be provided through a more comprehensive overview of Goffman's work on adaptions to prison life and its unauthorised underlife. This is followed by a section on data from research in which I embed ideas on the geographies of carceral object mobility in today's prison practise and make a case for including these micro-worlds in wider carceral geographic debates.

Object mobility in a carceral context

Whilst mobility is by now an established area of interest in carceral geography (Gill, 2013; Mountz, 2013; Philo, 2014; and others), objects have previously been mentioned in passing in a carceral context (see Philo, 2001, on Kantrowitz, 1996). Drawing on OOO, wider human geography has made an advance at questions like why objects are inherently political and related to power structures (Jackson, 2000; Shaw and Meehan, 2013). Research on matter and materiality has focused on the theory of human access to objects, but Meehan et al. (2013) contend that a theory of objects beyond this platform is needed. Focussing on aspects that make them more-than-human (see also Bennett, 2004), Latour's (2005) take on what makes up the 'social' and reassembling social connections of subjects and objects within an Actor-Network-Theory (ANT)[3] approach builds on the inherent agency of both[4]. Conceptualising objects means to understand objects not just as 'things' but as entities that 'harbor an erupting infernal universe within' (Harman, 2005: 95; Bennett, 2004). This 'more-than-is-visible' edge to objects is described by Bennett (2004: 351) as 'thing-power'; and by Meehan et al. (2013: 61) as their 'worldliness'. Contending that objects are 'force-full' and 'brimming with affect, productive of difference and generative of power', Shaw and Meehan (2013: 220) put them on the agenda of security geographies. OOO has engaged with questions of mobility and stillness, and has thereby re-imagined geographic terms like space and place (such as in the session 'The difference of things' by Dixon, Marston and Woodward at the 2011 Annual Meeting of the Association of American Geographers).

In order to gain a better grasp on the connections between relations that are built up between people and objects, ORT and psychoanalytic geographies (see Callard, 2003; Philo and Parr, 2003) distinguish between internal objects and the interaction between inner and outer factors and their effect on identity and self (Cashdan, 1988; Sayers, 2000). ORT understands the object as an entity that resists and at the same time enables representation questioning the ability of humans to rationally separate out epistemological and emotional life (Phillips and Stonebridge, 1998). Winnicott (1971: 1986) argues that people 'use objects and spaces to transition between the inner world of psychic fantasy and the outer world of objective reality' (in Shaw, 2010: 792).[5] This psychoanalytic approach to 'transitional' space offers a new dimension to the object-oriented philosophy that positions objects as inherently political and relational beyond the distinctly 'human'/'nonhuman' binary.

In everyday prison practice, objects are mainly regarded as either mundane or as security concerns and subject to vigilance. Object mobility comes into play in relation to pathways of contraband or other breaches of security. It is not just these highly charged objects, though, that are worth researching but rather a wealth of valuable findings can be drawn from considering everyday things. Authorised and unauthorised objects, and their moving pathways around confined spaces, make a difference to prisoners' sense of self and identity formation in relation to the institutionally imposed perception of 'prisoner' as well as an internally formed version of one's own self. Moving objects majorly contribute to what Ferrant (1997) terms 'symbolic spaces' in prisons, ones holding layers of unique and heightened meanings. Prisoners thus try to carve out personal space by imbuing material spaces and mobile objects with intangible meanings that express their identity or respective status within a particular prison space (Ferrant, 1997). Baer (2005) applies de Certeau's (1984) concept of 'tactics' to understand the importance of objects and to make sense of how they are used in confined spaces. He uses the example of a radio to underline his argument that it is only an object in the corner of the room until it is used/turned on, when suddenly it transforms the space and embodied experiences within – objects thus have an effect (de Certeau, 1984: 211). This visual, audible and haptic imprint on the interior of the prison is reflected in objects that are, in turn, intimately connected to people's tactics and their social words. The alteration of prison space through a tactical use of objects means that prisoners transform spaces of confinement into (more) personal space (Baer, 2005: 215), but they also display a prisoner's status and maintain and/or build a certain identity through objects. Other carceral geographers mention objects that are crucial to prisoners' sense of self and their struggle to maintain their identity through the manifold transitions into and out of prison environments, like mobile phones, computers (Moran, 2012: 572), or lipstick, particularly noting the gendered dimensions of objects and disciplinary normalisation (Moran et al., 2009).

The practice and performance of object mobility is entwined with a complex array of carceral particularities. Whereas geographies of mobility are often concerned with different forms of human movement (e.g., Cresswell and Merriman, 2011), these micro-scale mobilities play out in an environment of restricted mobility that partly rests on hidden mobility and practices of disguise. As Urry (2007: 7) points out, mobility 'is a property of things and people', and moving 'can be a source of status and power' (Urry, 2007: 9) whereas the coercion of movement is linked to social deprivation in exclusion. This is certainly a common denominator of prison environments, which augment questions of the connection between inhibition of movement and boundaries. Urry (2007: 290) highlights how 'securing' mobilities means 'securing' people within 'multiple panoptic environments', which seems to suggest the spreading of disciplinary power into society overall. Mobilities on the small scale of the carceral, however, reveal a complex picture where mobile and inertial practices can both stand for the display of status and power as well as being a diagnostic of deprivation and disempowerment. This matter has been pointed out on a larger scale of carceral movements by Gill (2013) and Michalon (2013), who question the association between mobility and freedom, and propose that much confinement can be found in spaces of mobility.

To capture these different elements, it is worth considering people's tactics as a merging of psychoanalytical-geographical conceptualisations like ORT with ANT-related takes on reassembling the social (Harman, 2014; Latour, 2005); thus underlining the connection between inner and outer objects and their relationality to human behaviour. To illustrate the intricate power relations between objects and prisoners, I will now turn to Goffman's (1991) work on secondary adjustments and the multifaceted dynamics of object mobility behind bars. The potentially serious theoretical discontinuities of bringing together thoughts on objects, mobilities, and their relationality will be unravelled with the help of Goffman's micro-scale conceptualisations of institutional life and underlife.

On the geography of 'floaters' and Goffman's secondary adjustments

Erving Goffman's sociology and his work on so-called total institutions uses a bottom-up approach concerned with individuals in specific locations and their inter-personal relations. While his concepts have met with severe criticism within carceral geography (Baer and Ravneberg, 2008; Moran, 2014; Moran and Keinänen, 2012), his work on total institutions helps to explain specific carceral characteristics with an in-depth take on inner dynamics and social encounters, but also on what he terms 'geographies of license'. Goffman describes the 'total institution' as a 'social hybrid', 'part residential community, part formal organisation' (Goffman, 1991: 22). He notes that 'total institutions' are part of the overall social order in society, with a 'fluid' but clearly tangible distinction between 'inside' and 'outside'. While every organisation involves a discipline of activity, Goffman establishes that every organisation also involves a 'discipline of being – an obligation to be of a given character and to dwell in a given world' (Goffman, 1991: 171). The institutional underlife and what he terms 'secondary adjustments'[6] therefore thrive as a kind of 'absenteeism' that originates in deviating from the prescribed way of being in such an organisation. Secondary adjustments recall Foucault's (2007) elaboration of *contre-conduite* (concepts of governmentality) and individual examples of 'counter conduct' or contestation.[7]

Goffman separates two different levels of secondary adjustments in what he calls 'make-dos' and 'working the system'. Make-dos do not require much involvement in and knowledge of the institutional system, meaning that people use available items in unintended ways thus modifying the institutionally predetermined conditions of life. 'Working the system', meanwhile means that the actor must have intimate knowledge of the system and its rules, utilising such knowledge to improve living conditions in a fashion that could be likened to 'tactics' (see de Certeau, 1984). Interestingly for geographers, Goffman dedicates a whole sub-chapter to 'places' and the question of the setting with regards to underlife activities. He separates institutional space into three categories: (1) space that is off-limits or out of bounds; (2) surveillance space; (3) a third space ruled by less than usual staff authority (1991: 204). Particularly focussing on the latter, Goffman describes how these regions which he calls 'free places' are characterised by bounded physical space that experience reduced levels of restriction and surveillance, 'where the inmate could openly engage in a range of tabooed activities with some degree of security' (Goffman, 1991: 205): 'License, in short, has a geography' (ibid.).

Goffman analyses this 'geography' of free places that function as a 'backstage' to the usual performance of staff-inmate relationships, already initialising his later research on performance and performativity. Free places are transient and fluid (reminiscent of Foucault's heterotopias; see Schliehe, forthcoming) and often connected to ever-changing inmate structures and territories (1991: 213). Territory formation, intimately connected to identity and self-formation, is linked to space in which one has 'some margin of control' similar to the use of objects (Goffman, 1991: 219). Considering objects and how they are usually handled in everyday life, Goffman explains that storing them for safekeeping in certain places like boxes and cabinets holds a certain meaning beyond the initial object: 'these places can represent an extension of the self and its autonomy, becoming more important as the individual foregoes other repositories of selfhood' (Goffman, 1991: 220f.). If it is difficult or forbidden to keep possessions and objects are used by others, so the inmate finds little protection from social contamination. The things that are controlled and have to be given up might be those that hold a special sense of self-identification. Personal storage space that conceals objects to keep illegitimate attention and legitimate authority at bay can avail complex structures. In closed institutions private accessible storage places are rare. Personal items are often stored beyond the inmate's reach, money is kept in administration and valuables or breakables are locked up for safety. Many of these objects can be an integral part of the body image, and items such as cosmetics needed to present oneself in a certain way might be collectivised or made available only to certain inmates or at certain times (Goffman, 1991: 222). Goffman describes it as a characteristic of institutional life to carry objects on one's own person, developing their own object mobility. This form of mobility is termed a 'transportation system' by Goffman, one which works with fixed and portable 'stashes' that are often found in free places and territories: 'if secondary adjustments are to be effectively worked out, an unofficial, usually undercover, means of conveying relevant objects has to be established' (Goffman, 1991: 225). This transportation system is employed as part of the institutional underlife that moves bodies, artefacts or objects and written or verbal messages.

One example employed by Goffman is books known as 'floaters':

> These books ... pursue a furtive, underground existence, rather like crooks on the run. They are passed from hand to hand, under the cover of shirts or jackets. They fly mysteriously into one's cell as the landing orderly is passing; they creep beneath the tables at mealtimes; they hide on top of the cistern in the recess. And, in the event of a surprise turnover, they frequently leap precipitously from cell windows rather than face discovery and arrest. (Goffman, 1991: 227)

These objects can be mobilised through relatively institutionalised collective transportation practices as well as through more individual routes. As important to Goffman's analysis as the circulation of objects is the mobility of messages and knowledges (1991: 228). Different types of communication – face to face, by messages, gestures and particular use of language, as well as the institutional development of mediated systems – can be employed for unauthorised transmission. The mobility of objects, persons or messages through similar systems shows how routes of transmission can be used for different items from minor rule-breaking to strongly tabooed contraband (Goffman, 1991: 231).

Considering social structures in closed institutions, Goffman describes many complex routes of social control that are used for secondary adjustments. One way is 'unrationalised force' or private coercion that is commonly found in many different forms (Goffman, 1991: 234). Another is economic exchange involving trade of objects or favours, one of the more widespread involving tobacco (snout) or lighters, and others might be selling of liquor or other contraband. Selling or bartering prizes won at bingo (Goffman, 1991: 238) is common, as well as other mediums of exchange like cigarettes. Gambling is a common undercover source of goods. Sale, barter, and exchange are elements of social organisation through which 'use-of-the-other occurred among inmates' in an important but unofficial way (Goffman, 1991: 242). There is another important mode of exchange that Goffman calls 'social exchange', through which signs and symbols of concern for another are reciprocated in a two-way transfer. This form of social payment is used as a measure of appreciation or a relationship (Goffman, 1991: 244).

Social exchanges – sustained by all different types of relationships – are characterised by little resources available in the 'reduced circumstances' of institutional living, which is why some of the secondary adjustments are practised to develop or deliver 'ritual supplies' (e.g., Valentine or Christmas cards but also food, tobacco, and other objects) (Goffman, 1991: 247f.). The exchange of these ritual supplies happens between inmates, inmates and staff, as well as inmates and visitors, proving the semipermeable nature of the institution (Goffman, 1991: 247f.). Goffman observes intricate chains of contact designed to transport and deliver ritual supplies (Goffman, 1991: 251). This form of thwarting authoritative structures, according to Goffman means psychological survival in which information about and knowledge of the system become crucial goods. In this culture, then, 'wising each other up' becomes part of inmate relationships (Goffman, 1991: 252).

Goffman explains that the three arrangements through which one individual can make use of objects or the services of another (private coercion, economic exchange, and social exchange) are all simplified analytical structures. In practice, several of them are often simultaneously and routinely used (Goffman, 1991: 258). Thus, a description of institutional underlife provides a 'biased picture' of life in it. Depending on the degree that inmates stick to primary adjustments (i.e., simply to the formal/overt rules), underlife may be unrepresented or even unimportant (Goffman, 1991: 262). In any case, it is important to observe how a particular institution is worked and how institutions in general might be worked; analysing social control and bond formation makes this easier. Detainees' agency and their effort to build an identity acquiescing separate from the institutional doctrine is mainly played out in the spaces of secondary adjustments, thus creating a (limited) space for agency and the (re-) imagination of self through the mobility of objects.

Objects and their carceral mobilities as extensions of self

The following data presented is part of an extensive research project that involves young female detainees in different types of closed institutions in Scotland, as well as their staff. The project has a broader focus on young women's experience of different closed institutions like secure care units or psychiatric facilities (see Schliehe, 2014, forthcoming), and their embodied experiences of confinement. Both the conceptual and empirical material reflect issues of disciplinary power and the constitution of institutional life. For this chapter only prison interviews were considered, comprising of 24 interviews with young female prisoners and 12 prison staff.[8]

Table 9.1 Sheet of allowed possessions (PR, field notes, 07/14)

Prison Uniform:

Polo Shirts (3)

Sweat Shirts (2)

Jog Bottoms (2)

Shoes/Boots (1)

House Coat (1)

Pyjamas (1)

Jacket (1)

Personal Clothing:

Earrings (1 small)

Watch (1)

Ring (no raised face or stone)

Dressing Gown (1)

Slippers (1)

Pyjamas (1)

Tops (4)

Bottoms (2)

Shorts (1)

Joggers (gym) (1)

Shoes (3)

Ladyshave

CD Player

CDs (max 5 pre-recorded)

Underwear/socks

Photographs (max 20)

When arriving in prison, detainees' are confronted with a new space and their allotted cell's condition is recorded, and the prisoner is made aware that it is her job for keeping the cell in good condition early on (field notes, 09/14). The 'cell checklist' requires prisoners to think about the objects in their living environment, assuming responsibility for the object's state (e.g., maintaining and not damaging things like chairs, blinds, kettle, mirror, and bin). The same happens with prisoners' own possessions. Initial processing involves leaving most possessions like jewellery, phones, purse, and some personal items like hair extensions and clothes in storage and out of a prisoner's reach. Prisoners' details are taken and they are given their 'jail number'. Before being moved to their block, they are strip-searched to check for contraband, which sets the institutional scene of allowed and forbidden objects. Every prisoner receives a starter pack of clothes that represents the mundane items in everyday life. The so-called jail uniform is worn by every prisoner in the study prison, but, depending on status, enhanced[9] prisoners are allowed more personal clothes, shoes, and accessories. Everyone is limited, though, in the number of items allowed and the times and circumstances when they are authorised to be worn. During the day every prisoner must wear what they call their 'uniform' but many find small tactics to thwart the rules related to clothing (make-do clothes adaption). It depends on connections outside – to family members or others – how much and if any personal objects are sent in for a prisoner. An object's mobility thus underlines the semipermeable nature of closed spaces, while being clear on the institution's role of having the upper (disciplinary) hand. Most prisoners know the exact number of items allowed to them (see Table 9.1). While some prefer to wear prison clothes for comfort or not to 'waste' own items, others maintain their self-image by wearing their own clothes:

> I walk about wi' the best of stuff. And I wouldnae change that. I don't dae it to show aff ... I just dae it because that is me ... when you get your own material in, like, and you're dressing the way you do outside you feel cleaner and you feel a wee bit better about yourself, obviously. Aye, you do, definitely. Compared to wearing jail ... aye, you feel more in reality. (Sophia)

Prison clothes are reported to influence how prisoners feel about themselves. Clothing and styling carry a meaning that is not just important for a prisoner's individuality and for their own image of self, but also reflects the inner hierarchies and workings of the block collective. Adaptations to clothing beyond prison regulations show how prisoners deviate from prescribed being and thus symbolise some fragments of autonomy.

When being caught breaking rules, a prisoners' status is often downgraded, setting the scene for object removal. This involves having to pack up all belongings that are no longer allowed in black plastic bags and handing them back to 'reception', where they are stored. The open display of downgraded objects being bagged and dragged down corridors reveals the loss of status to everyone else (field notes, 07/14). It is not just clothes that make up prisoners' status but also objects like cell decorations, the display of shampoo bottles, perfumes, or air fresheners as well as putting up pictures and cards that are sent in from people outside and from other prisoners. Seeing these objects as an extension of self and one's own autonomy means that much more than just an object is removed when downgraded.

Despite the threat of downgrade or segregation, prisoners regularly try to sidestep or bend the strict rules in confinement. There are constant smaller transgressions (two-way exchange/trade of objects) like the swapping of supplies where older prisoners sometimes have agreements with younger ones to buy them tobacco and supply them in return for other items on the 'shop sheet' with equal value. Others play games involving imaginary mobility and imaginary 'forbidden' items without actually possessing them; like Emma who says, 'I go on holiday and I will take ...' remembering different sorts of alcohol, like Buckfast or Frosty Jack. Another favourite is gambling:

> It's not even like you allowed games and people have blown that with gambling. You can't gamble much in there. It's not as if you've got money but people still do. I don't know, with coffee or toiletries or something like that (Kirstie).

Contraband like drugs get 'in' on occasion and people have been said to smoke cannabis – trying to mask the smell with air freshener or smoking at night when staff could not find out. 'Free places' in this case are often cells that are used in transient and fluid ways – moving objects around the block – but at the same time providing limited levels of reduced surveillance. It is a rare occasion, though, when prisoners manage to smuggle in medication that can be used as an 'upper' or 'downer' ('you cannae get oot your nut unless somebody holds their med-I [medication] back, know what I mean? And you cannae really dae that often' (Bianca)). During the day it is difficult to evade the staff gaze and hide transgressions; prisoners do not seem to try this often (or are not found out and do not talk about it). At night, more possibilities for such transgressions present themselves as the spaces are not as heavily controlled, but the radius of object movement is correspondingly a lot smaller; prisoners only get out of the room for a toilet trip of seven minutes, one at a time. Prisoners do, however, leave items at other people's doors:

> You can go to their [other people's] doors and that. Obviously open the latch and see what's happening. 'Want to leave something at my door?' Things like that. And then obviously you go oot and they can put something at your door and you can go oot [out] and get it which is alright that way. (Charlotte)

Geographies of license take many forms and can be transformed from everyday objects to transgressive entities through their mobility; for example, some prisoners, as Louise told me, throw lines out of their windows to pass things around to share objects. Others report that everyday items are used to bully or degrade other prisoner, notably by sticking sanitary pads or toilet paper on certain cell doors at night. Even the sheer threat of particular items can have a detrimental effect on prisoners and their status (as a form of private coercion), leading to social control practices among prisoners as well:

> [I got sent to segregation] but it gets called the Digger <sigh> ... people making accusations, saying that I had weapons in my room when I didnae, so I got

put doon there under investigation and I came back up and I got a three month downgrade and a 7 'all round',[10] but I had my room searched and they never found nothing, so where's their evidence? I still got punished. (Isla)

Actual and imagined object mobility holds keys to control and self-assertion, as well as symbolising powerlessness and inertia. Much movement of objects occurs in line with the movement of people within each block and the wider prison architecture and regime. Objects move from cell to cell while prisoners are unlocked. They are dropped into the staff office or into the dining room at meal times, and are moved to and from the gym or library at set times during recreation. At lock-up, prisoners quickly have to move to their own cell, which then restricts object mobility to the confined space of the cell. There is an almost constant mobility of news and exchange of social interaction through locked doors, walls, and windows. Wising each other up and learning to work the system is an essential part of prison (under)life. As Emily explained, objects like (ritual supply) cards find their way to other parts of the prison, other prisons and sometimes even people in segregation who are living deprived of most objects, all through complicated transportation systems beyond the researcher's reach. There are many more objects 'floating' around prison (food, cleaning products, razors, shopping, and even rubbish) that are used in many small ways to re-establish control and certain levels of autonomy. The symbolic and emotional effect of objects is indeed palpable when delving into the micro-worlds of secondary adjustments. The multiple and complex mobilities of objects reach beyond the cell, the block and prison circumference into other people's homes and storage cupboards. Losing, leaving, and retrieving objects inside and outside prison walls affects one's own sense of self and the constant struggle with prescribed, normalised, and resistant being as Goffman uncovered in relation to institutional adjustments.

Conclusion

The above observations and findings underscore the importance of micro-scale processes for understanding embodied experiences of carceral space. While many object movements are inherently embedded in the prison regime – from the route (planned movements of prisoners' bodies around confined space, e.g., from workshop to block) to distribution of food, laundry, or shopping – just as much seems to happen undercover in secondary adjustments integral to intricate social interaction and identity formation. This chapter points towards the pivotal role of objects and their movement to understand the constitution of carceral space and regime. Productively unpacking 'floating' objects as a particular form of constrained locomotion is key to bringing carceral objects into the analytical focus. More conceptual work is needed to work through and consider the potential of OOO/ORT/ANT and other theoretical concepts, and more in-depth lines of enquiry could be developed to consider how, for example, gender and sexuality shape object mobility.

While a focus on mobility is on the carceral geographic agenda with an understanding of carceral space 'involving constant circulation, porous borders and unruly time' (Armstrong, 2015: 2), it is nonetheless still a largely neglected dimension of imprisonment. The importance of considering objects presents a further under-researched area of geographic research on confinement. Conceptually involving carceral research in object-related philosophy and connecting to debates on psychoanalytical geographies would be an important future task. There is scope here to develop strong empirical and conceptual arguments around object mobility. The empirical material shows, on the one hand, how the constrained locomotion of objects mirrors the movement of and restraints on bodies, but also, on the other hand, brings out the complex dynamics of institutional underlife and prisoners' secondary adjustments. Object mobility exposes much beyond the obvious: symbolising how prisoners work the system and exercise control, but also disclosing through contested practices of storage, how prisoners' identity and status become such powerful markers of control over bodies and minds. Generally, the underlife of the institution is practised in myriad small ways of trying to resist, even to undermine institutional as well as interpersonal controls. While this challenges a simplified sense of institutional power imposed upon prisoners, revealing instead agency, it also captures the regimented mechanisms of securitisation and coercion working within closed spaces. If nothing else, the detailed geographies of floating objects reveal another layer of the complexity and opacity of confinement.

Acknowledgements

I would like to thank Ian Shaw for his help with the conceptual frame for this paper, and Chris Philo and Hester Parr for comments on an earlier version of the chapter!

Notes

1 Locomotion is described as the action or power of movement from one *place* or position to another and the capacity for this (OED 2016). The added emphasis on a particular place as well as the inherent capacity needed for moving set an intriguing 'terminological' background for object research in prisons as these have particular connotations and importance there.
2 According to the OED (2016) 'floating' is defined as 'not fixed or settled in a definite state or place; fluctuating, variable, unstable' and 'having little, or comparatively no attachment'. This state of disconnection and instability is helpful when thinking through prison life and the social world of moving objects within it. Taking up a Goffmanesque term (see below), 'floating' like 'drifting' (see Peters, 2015) has not been sufficiently unpacked within a mobilities framework. The inherent transgressiveness of floating e.g. when failing to adhere to authorized channels or 'conduits in space' (Peters, 2015: 270) offers important conceptual cues for analysing objects, confined spaces and institutional underlife.
3 There is a lot to be said about ANT's attentiveness to 'objects' and 'things' from trains to hinges as a way marker for 'non-human-geographies' (e.g., Laurier and Philo, 1999).
4 See Law (2002) for putting the concept of fluidity back into object networks.

5 Aitken and Herman (1997) argue that D. W. Winnicott's notion of transitional space and his emphasis on play together provide a refreshing outlook on the processes identity formation. The principal concerns being with how children (and adults) bridge the gap between egocentricism and recognition of an external world (and objects), and how they negotiate and renegotiate the relations between self and other.

6 Primary and secondary adjustments to the institution, comprise what Goffman names the 'underlife of the institution', dividing them into 'disruptive' [attempting major structural changes and rupture] and 'contained' [fitting into existing institutional structures] secondary adjustments (Goffman, 1991: 180). Here, only inmates' secondary adjustments will be considered, although Goffman describes both staff and inmate adjustments. Goffman uses the term secondary adjustment but at the same time calls it into question as 'clumsy' and dependent on the point of view, as psychiatric doctrine does not accept secondary adjustments possible for patients due to their definition as mentally ill (Goffman, 1991: 186).

7 Foucault (2007) discusses *conduites* and *contre-conduites* in the March 1, 1978, lecture of *Sécurité, territoire, population*. In his lecture series Foucault mentions historical examples in the context of his genealogical discussion of societal transformation through bio-power (see also Davidson, 2011; Death, 2010).

8 Field diary entries are filed by month and year whereas transcripts of interviews are cited by prisoner name which they could chose for themselves at the start of the interview. For reasons of confidentiality all references to the specific prison site and time of interview (and specific day for field diary entries), as well as information like specific age, are not included.

9 There are three levels of status for prisoners. They come in as 'standard' and can progress to 'enhanced' which comes with more privileges and access to objects; but they can also be downgraded to 'basic', which in turn comes with loss of privileges and removal of items.

10 Downgrade of prisoner status involving losing most possessions as well as 7 'all-round', which means a week without TV, recreation, or pay.

References

Armstrong S (2015) The cell and the corridor: Imprisonment as waiting, and waiting as mobile. *Time and Society* 00: 1–22.

Baer LD (2005) Visual imprints on the prison landscape: A study on the decorations in prison cells. *Tijdschrift voor Economische en Sociale Geografie* 96(2): 209–217.

Baer LD and Ravneberg B (2008) The outside and inside in Norwegian and English prisons. *Geografiska Annaler: Series B, Human Geography* 90(2): 205–216.

Bennett J (2004) The force of things: Steps towards an ecology of matter. *Political Theory* 32(3): 347–372.

Callard F (2003) The taming of psychoanalysis in geography. *Social and Cultural Geography* 4(3): 295–312.

Cashdan S (1988) *Object Relations Therapy: Using the Relationship.* New York, NY: WW Norton & Company.

de Certeau M (1984) *The Practice of Everyday Life.* London: University of California Press.

Cresswell T and Merriman P (2011) *Geographies of Mobilities: Practices, Spaces, Subjects.* Farnham: Ashgate.

Davidson AI (2011) In praise of counter conduct. *History of the Human Sciences* 24(4): 44–51.

Death C (2010) Counter-conducts: A Foucauldian analytics of protest. *Social Movement Studies* 9(3): 235–251.

Ferrant A (1997) Containing the crisis: Spatial strategies and the Scottish prison system. Unpublished doctoral dissertation, University of Edinburgh.

Foucault M (2007) *Security, Territory, Population: Lectures at the College de France 1977–1978*. Basingstoke: Macmillan.

Gill N (2013) Mobility versus liberty? The punitive uses of movement within and outside carceral environments. In: Moran D, Gill N and Conlon D (eds) *Carceral Spaces: Mobility and Agency in Imprisonment and Migrant Detention*. Farnham: Ashgate, 19–36.

Goffman E ([1961] 1991) *Asylums: Essays on the Social Situation of Mental Patients and Other Inmates*. London: Penguin Books.

Harman G (2005) *Guerilla Metaphysics: Phenomenology and the Carpentry of Things*. Chicago, IL: Open Court.

Harman G (2014) *Bruno Latour Reassembling the Political*. London: Pluto Press.

Jackson P (2000) Rematerializing social and cultural geography. *Social and Cultural Geography* 1(1): 9–14.

Kantrowitz N (1996) *Close Control: Managing a Maximum Security Prison-The Story of Ragen's Stateville Penitentiary*. New York, NJ: Criminal Justice Press.

Laurier E and Philo C (1999) X-Morphising: Review essay of Bruno Latour's Aramis, or the love of technology. *Environment and Planning A* 31(6): 1047–1071.

Latour B (2005) *Reassembling the Social: An Introduction to Actor-Network-Theory*. Oxford: Oxford University Press.

Law J (2002) Objects and space. *Theory, Culture and Society* 19(5/6): 91–105.

Meehan K, Shaw IGR and Marston SA (2013) Political geographies of the object. *Political Geography* 33: 1–10.

Michalon B (2013) Mobility and power in detention: The management of internal movement and governmental mobility in Romania. In: Moran D, Gill N and Conlon D (eds) *Carceral Spaces: Mobility and Agency in Imprisonment and Migrant Detention*. Farnham: Ashgate, 37–56.

Moran D (2012) Prisoner reintegration and the stigma of prison time inscribed on the body. *Punishment and Society* 14(5): 564–583.

Moran D (2014) Leaving behind the 'total institution'? Teeth, transcarceral spaces and (re) inscription of the formerly incarcerated body. *Gender, Place & Culture* 21(1): 35–51.

Moran D and Keinänen A (2012) The 'inside' and outside' of prisons: Carceral geography and home visits for prisoners in Finland. *Fennia* 190(2): 62–76.

Moran D, Pallot J and Piacentini L (2009) Lipstick, lace, and longing: Constructions of femininity inside a Russian prison. *Environment and Planning D: Society and Space* 27(4): 700–720.

Mountz A (2013) On mobilities and migrations. In: Moran D, Gill N and Conlon D (eds) *Carceral Spaces: Mobility and Agency in Imprisonment and Migrant Detention*. Farnham: Ashgate, 13–19.

OED (2016) *Oxford English Dictionary*. Oxford: Oxford University Press, *s. vv.* 'floating', 'locomotion'.

Peters K (2015) Drifting: Towards mobilities at sea. *Transactions of British Geographers* 40(2): 262–272.

Phillips J and Stonebridge L (1998) *Reading Melanie Klein*. Abingdon: Routledge.

Philo C (2001) Accumulating populations: Bodies, institutions and space. *International Journal of Population Geography* 7(6): 473–490.

Philo C (2014) 'One must eliminate the effects of … diffuse circulation [and] their unstable and dangerous coagulation': Foucault and beyond the stopping of mobilities. *Mobilities* 9(4): 493–511.

Philo C and Parr H (2003) Introducing psychoanalytic geographies. *Social and Cultural Geography* 4(3): 283–293.

Sayers J (2000) *Kleinians: Psychoanalysis Inside Out*. Cambridge: Polity Press.

Schliehe AK (2014) Inside 'the carceral': Girls and young women in the Scottish criminal justice system. *Scottish Geographical Journal* 130(2): 71–85.

Schliehe AK (forthcoming) Re-discovering Goffman – contemporary carceral geography, the 'total' institution and notes on heterotopia. *Geografiska Annaler Series B Human Geography*.

Shaw IGR (2010) Playing war. *Social and Cultural Geography* 11(8): 789–803.

Shaw IGR (2012) Towards an evental geography. *Progress in Human Geography* 36(5): 613–627.

Shaw IGR and Meehan K (2013) Force-full: Power, politics and object-oriented philosophy. *Area* 45(2): 216–22.

Urry J (2007) *Mobilities*. Cambridge: Polity Press.

Whatmore S (2006) Materialist returns: Practising cultural geography in and for a more-than-human world. *Cultural Geographies* 13(4): 600–609.

Winnicott DW (1971) *Playing and Reality*. London: Travistock Publications.

Winnicott DW (1986) *Home Is Where We Start From: Essays by a Psychoanalyst*. Harmondsworth: Penguin Books.

Part III

Distribution

v. the way in which things are shared out, spread, or supplied

10 Virtual presence as a challenge to immobility

Examining the potential of an online anti-detention campaign

Emma Marshall, Patrycja Pinkowska, and Nick Gill

Introduction

Previous work by geographers has used the framework of mobilities, immobilities, and moorings (Hannam et al., 2006) to examine how the expanding use of detention may be employed by the state to exclude and disempower those migrants it seeks to deport (Gill, 2013; Mountz, 2011; Mountz et al., 2012). Activists and anti-detention campaigners in the UK have also recognised the often geographically remote locations of Immigration Removal Centres (IRCs) as a factor contributing to the silencing of those who are detained, thus enabling the state to keep detention largely 'out of sight, out of mind'; arguably making the denial of basic rights, and limited access to justice within detention, pervasive (Bail for Immigration Detainees, 2009). In response to these academic and activist concerns that detention is intentionally situated away from public view, we evaluate the opportunities that the Internet might offer as a challenge to the physical isolation of detained populations. We consider the extent to which alternative, and multiple, narratives of those held in detention facilities may be mobilised via online anti-detention campaigns, focusing on the capacity of the Internet to enable a virtual presence; which we assert can go beyond the physical boundaries of detention to make experiences of detention visible, as a distinct form of political action.

Georg Simmel's exploration of different forms of presence (Simmel et al., 1997; Gill, 2013; see also Gacek, this volume) has been used in recent research to develop fluid understandings of mobility that go beyond the corporeal properties of people and objects (Büscher and Urry, 2009; Urry, 2007). This rethinking of 'presence' beyond the physical subject is the basis for new 'mobile methods', which seek to examine the relationships between 'proximity, distance and movement' (Urry, 2007: 20) of people, images, information, and objects (Büscher and Urry, 2009; Fincham et al., 2010; Urry, 2007), including tracing the 'virtual mobilities of people' through online sources (Urry, 2007: 40). Using a broad conceptualisation of presence has important implications for people detained in IRCs, as although their physical movement is restricted, it recognises that other forms of their presence may become mobilised. Previous academic research has investigated how the collective identities of detainees have become mobilised through public opinion and the popular press, often having a negative impact on processes of political decision-making (Bosworth and Guild, 2008; Mountz et al., 2012; Tyler, 2010). In this chapter we offer an alternative account, by investigating how new forms of collective presence mobilised through online campaigns may offer significant potential for political action. We explore how virtual presence may be appropriated by online campaign groups, and used productively as a political act of empowerment or resistance, by opening up spaces of detention to the general public, and bringing previously hidden voices into the public domain.

In order to explore the potential of virtual mobilities for political activism, we evaluate the creative tactics (de Certeau, 1984; Gill et al., 2014) employed by the Detention Forum, a network of 37 voluntary sector and charitable organisations that support, and campaign on behalf of, those who are detained, working together to challenge the use of detention in the UK (The Detention Forum, 2016). The Detention Forum is hosted by the Refugee Council, a national charity, and is overseen by a Co-ordination Group made up of representatives of a variety of member organisations in the network. We have chosen to focus on the Detention Forum's online campaign 'Unlocking Detention' (referred to from here as 'Unlocked'), because it has been developed to provide people with a way of sharing their experiences of detention, whether these come from people who are, or have previously been, detained, or from friends, family, or the volunteers who make regular visits to IRCs. Unlocked has its own website, which includes blog posts, a link to the campaign's Twitter feed, basic facts about detention, and opportunities for the public to become involved in the campaign themselves (Unlocked, 2015). People who are detained may not have direct access to the campaign on social media, so visitors to detention help to publicise the campaign to people within IRCs, relaying information and personal accounts of the conditions inside detention back to the website.

Drawing on research material gathered through analysis of the Unlocked blog posts and the Twitter feed, and a semi-structured interview conducted with two of the Coordination Group members, we evaluate how making experiences of detention visible to the public can be used to assert a collective voice to challenge mainstream narratives about detention in the UK. We use narrative analysis (Eastmond, 2007) to explore how recurrent themes, through personal accounts of detention online, can connect the reader to the sites of IRCs, highlighting the importance that migrant narratives have for exploring the complexities of identity in relation to mobility (Valentine et al., 2009), and the expression of these narratives as a means of political resistance (Smith, 2015; Wooley, 2014).

Mainstream narratives of detention

Immigration detention, defined as a place where non-citizens are kept for the purpose of immigration related goals (Silverman, 2012: 1134), is considered an administrative and non-punitive measure, and is generally permitted in international law (Wilsher, 2008: 897). Expanding since the 1990s, the detention complex is by now firmly rooted in Western democracies' strategies for the control of migration. There are currently eleven IRCs in the UK, where around 30,000 migrants are detained every year.[1] The institution of detention, and its unprecedented expansion in liberal, democratic states, forces us to consider the many contemporary international, legal, political, economic, and moral issues that have provoked growing debate amongst activists and academics from several disciplines. Geographers are contributors to those debates, stressing not only the political urgency of this scholarship (Mountz et al. 2012: 524), but also pointing out the spatial state strategies of confinement and control that govern detention practices (Gill, 2009, 2016; Martin and Mitchelson, 2009).

Over the last couple of years, broadsheet British newspapers such as the *Guardian* and the *Independent* have provided critical journalistic enquiry into the UK detention estate. This has included coverage such as: investigations into the tragic deaths that have occurred at IRCs (*Guardian*, 2014); scrutiny of the private security firms that are contracted to run the IRCs (*Independent*, 2014); and high profile cases of unlawful detention, and compensation subsequently paid out by the UK government (*Guardian*, 2015). Furthermore, an undercover investigation by *Channel 4* in March 2015 captured the attention of the UK audience when it broadcast footage from inside one IRC, Yarl's Wood, showing members of staff, employed by Serco, the private security firm contracted to run it, uttering racist, abusive, and dehumanising comments (*Channel 4*, 2015). These undercover investigations by journalists have provoked public and political outrage; generating calls to review the use of detention. This has included backlash against the incarceration of perceived vulnerable populations such as women, as well as questions over systemic racial abuse, and the need to urgently investigate the conduct of the private security firms. It is important to note, however, that when such incidents of severe mistreatment of people in detention make it to the mainstream news, they often fail to provoke a debate that would call into question the legitimacy of the institution as a whole. The 'regrettable but necessary' narrative, so often used by the UK government (Silverman, 2012), still dominates both policy and public discourse.

Bosworth and Guild (2008) have noted that as detention facilities have been increasingly justified and expanded as a 'necessary' security measure, the negative identity of asylum seekers in the UK media has been reinforced, effectively erasing the distinction between the categories of 'asylum seeker' and 'criminal' (Bosworth and Guild, 2008: 710).[2] This is reflected by another type of media coverage focused on detention; the occurrence of hunger-strikes or riots. Melanie Griffiths (2014a) suggests this type of narrative is often preferred by the mainstream and local media; and when coverage of detention occasionally makes it to the news, it tends to be dominated by scenes of armed police and general chaos. This contributes to the view that male migrants, in particular, are a dangerous 'uncontrollable' population who need to be 'locked up' for public safety. These types of narrative generate an oversimplified version of detention, which become polarised along gendered lines vis-à-vis the image of the 'vulnerable detainee' versus the 'deserving detainee' (Griffiths, 2014b). In addition, questions arise about the legitimacy of detention in terms of who should, or should not, be detained, rather than whether detention is ever ethically justifiable as a non-punitive measure, or the conditions under which it may be deemed acceptable. In order to shift the nature of the debate to the *use* of detention, rather than generalised perceptions of detainees, different types of reporting are needed. This is where we believe that the Internet can offer important opportunities for alternative narratives to be mobilised, in order to raise public awareness of the systemic use of detention.

Potential of online activism

The importance of the Internet for political action has been widely theorised in recent years, partly due to its role during the Arab Spring (Aouragh and Alexander, 2011; Butler, 2011), as well as its potential for human rights campaigns (Jensen and Jolly, 2014). In this literature, the Internet is often perceived as more democratic than mainstream media, not least because it affords the possibility of participation to a wide group of users, who do not necessarily need to be 'professionals' to send out a public message. Its widespread availability makes it possible to bypass the often narrow focus of the mainstream media's news coverage (Schaffer and Smith, 2014), making it possible for alternative narratives to become visible. For the Coordination Group members overseeing Unlocked, the attractiveness of the Internet rests on its ability to circumvent the mainstream media and thereby sidestep the tendency to only cover certain angles of immigration detention that entrench the narratives they are at pains to dispel.

Coddington and Mountz (2014) have emphasised the importance of the Internet for people detained in places difficult to access. In the context of their research on Australian offshore detention centres, they found that it was not only used by people within the centres to communicate with families and friends, but also to share information about the conditions at the centres, to publicise detainees' protests and strikes, and to inform the mainstream media about otherwise unreported issues of humanitarian concern. Whilst those detained were physically incarcerated, they were able to mobilise a political presence online via virtual networks, illustrating how considerations of the Internet may offer significant potential for explorations of distance and mobility, as well as for discussions of political action, resistance, and power relations. Thus, the Internet may be conceptualised as a medium for overcoming the physical boundaries of detention: a fast, efficient, and cheap way to communicate and share information both with individuals and with wider audiences.

Despite the potential of the Internet for political activism, we are, however, far from suggesting that it is a straightforward solution to challenging the exclusionary practices of detention, particularly as access to the Internet is to some extent restricted within IRCs in the UK. To reduce the isolation of people in detention facilities, and to enable communication, all detention facilities in the UK now have computer rooms with access to the Internet; but whilst the Home Office denies blocking the websites of non-governmental human rights organisations, it admits obstructing access to social media sites like Facebook or Twitter, as well as free communication software like Skype, in the name of public safety.[3] Furthermore, the mobile phones given to people in detention are not Internet enabled, and have no cameras. Whilst they may be invaluable for sustaining contact with the outside world, the interactions that they facilitate are limited to conversations between people who already know each other, such as family members, friends or legal advisers. As people held within detention facilities may be prevented from making public posts online, or making new contacts outside of detention, restrictions on the use of technology can be understood as one way in which presence in the public sphere is 'confiscated' by the state (Gill, 2013: 27).

IRCs in the UK occupy physical locations that are notoriously difficult to access; whether located in remote countryside without good public transport links, or on industrial estates in proximity to major airports, they are often far from detainees' networks of support, families, and lawyers. Therefore, during their time in detention, many people rely on visits from local volunteers, which is often where support for anti-detention campaigns begins to mobilise.[4] The limitations on the availability of the Internet mean that online movements cannot take place without an organised effort, sometimes in ways that are riskier than others. Through our own research in detention facilities we have learned that some people are able to overcome restrictions imposed on them by accessing both Facebook and Twitter using proxy servers. In research we carried out online, we came across a Twitter account of someone detained, who was making public posts until the day he was forcibly removed from the UK. Mobile phones with cameras have also been smuggled into the IRCs, and the filmed footage can be seen online, for example on the Standoff Films website (Standoff Films, 2015)[5]. These methods expose detainees to the risk of revealing aspects of their case that could be used against them in the legal system, and to the risk that they will be disciplined within the detention centre if they are caught. For these reasons, and due to the duty of care owed to detainees, the Detention Forum never encourages this type of subversion. They have managed to develop safer methods through which detainees can coordinate with visitors to access online communities, though, which rely on those who visit detention acting as intermediaries, taking messages from people who are detained and posting these on their behalf.

Whether online campaign groups are able to make their voices heard in the public sphere cannot be reduced to the availability of, or access to, the Internet; it relies on human effort (Papacharissi, 2002), and 'the capacity of social actors for mobilisation, organisation and self-representation' (Barnett, 2004: 195). Our research interest in the online campaign Unlocked is centred on the ways in which it co-creates virtual presence for those in IRCs, who are otherwise excluded from the public domain. We acknowledge that the notion of 'self-representation' is problematic in the context of Unlocked, because messages are mediated by the visitors that relay them. To be clear, the Home Office blocks access to social media sites from within detention, which means that it is necessary to facilitate the campaign from outside detention in order for social media platforms to operate as a legitimate mode of communication between detainees and the general public. However, this also means that control over the content of the campaign remains located outside detention rather than within it, and the process of co-creating the virtual space of Unlocked is situated within broader relations of power between those who are detained, and those who are not.

Notwithstanding these reservations, Simmel's insight that we all possess multiple forms of presence alerts us to the ways in which the Internet, as a communication technology, allows some individuals to mobilise aspects of their identity online. We believe that despite the limitations of the Internet, it need not be conceptualised only as a medium or a means to an end, but also a 'space of politics' in itself (Barnett, 2004: 190); actively shaped, and created, by those who participate in it. Our decision to interview the Coordination Group members comes from an acknowledgement that they facilitate input from a collective of individuals and campaign groups. According to them, Unlocked was imagined as an umbrella space for anyone to talk about the injustices of detention.

Here one of the Coordination Group members explains how the campaign is intentionally built around multiple voices:

> Unlocking Detention is not a testimony project and ... whilst it's about all people who are detained, it's not a hierarchy ... I think it's really important that we've got the words of the people who visit and maybe family members and even NGO workers who visit detention for the first time and are actually really affected by it, because the people that will campaign against detention aren't just the people who have been detained, and it needs to engage with people who come from different backgrounds. I think that's one of the better ways of doing it really, is you grab the voice of the people who know best about detention by being detained, but you also hear from other sides of the story as well. It's about different actors in society. (Coordination Group member)

In these terms, the Coordination Group member explains how the motivation of those running Unlocked is to facilitate differing, and multiple, accounts of detention. While the interviewee recognised these are always partial and power-laden, the Internet provides a platform to express these perspectives to enable a stronger political voice. Establishing an online presence allows the campaign to mobilise an alternative collective to counter mainstream narratives that focus on examples of severe mistreatment within detention facilities, or the portrayal of detention as necessary for the public good.

Co-creating virtual presence: What does it feel like to be detained?

In this section, we outline the findings of our empirical research to investigate how Unlocked uses creative tactics to mobilise virtual presence by moving the reader in and through spaces of detention; and moving detained voices out of, and beyond, the physical boundaries of IRCs. We argue that in distinct ways these tactics unsettle the boundaries of detention. By encouraging the reader of the blogs to 'feel' what it is like to be inside detention, we argue that the Unlocked campaign moves the reader through spaces of detention itself, destabilising its relational proximity to the broader community. In order to evaluate the ways in which

Unlocked mobilises an online presence for those in detention, we use analysis of the written narratives of detainee experiences to explore themes from the blogs posted on Unlocked. Recognising that 'stories cannot be seen as simply reflecting life as lived, but should be seen as creative constructions or interpretations of the past, generated in specific contexts of the present' (Eastmond, 2007: 250), we use narrative analysis as a framework for considering how the experiences of detainees represented in the blogs relate to the broader framework of literature on detention, by tracing the key themes and their interconnections with existing theoretical perspectives.

Whether online information is intended for public or private use has featured in academic debates about the ethical standards that should apply to web-based research methods (Kozinets, 2015; Madge, 2007). We have chosen to present our research about the Unlocked campaign because the information it disseminates is explicitly meant to be viewed by the public, in order to engage with as wide an audience as possible. We reference the names of those who have written the blogs, in order to acknowledge their authorship. However, we have carefully considered and applied the principle of anonymity wherever we feel it is appropriate, in order to conform to the usual academic standards of ethical practice for working with potentially vulnerable groups.

Moving through spaces of detention

Most IRCs in the UK occupy remote locations, where they are 'concealed' from public scrutiny (Tyler, 2010: 69). To challenge this, the Unlocked campaign invites the reader into the spaces of the IRCs, proclaiming on its website: 'you are cordially invited to the tour of the UK detention estate – on Twitter!' (Unlocked, 2015). As photographs inside the IRCs are prohibited, most of the online campaign relies on text to create an impression of detention in the reader's mind, supported by a few images of the outside of the centres. Relying upon the online content, the campaign uses the Internet to build a material connection between places that are physically separate. In this way, the campaign may be seen to harness what Simmel views as the power of 'conception' (Simmel et al., 1997: 171); the ability of the human mind to connect two physically distant locations, through an act of 'path-building' (Urry, 2007: 21). In recreating the conditions of detention, in virtual form, the online content connects the IRC with the reader.

Coordination Group members note how the descriptions of detention provided by those who have been detained, or those who have visited detention, can help to mobilise the reality of IRCs, as it helps 'to create some detail; some texture for people's imaginations' (Coordination Group member). The connection that the campaign establishes has a significant emotional dimension, as it attempts to capture the support of the reader. Empathy can be defined as an emotional relationship where 'one person imaginatively enters into the experiential world of an other' (Bondi, 2003: 71), which describes the type of leap the Unlocked campaign encourages the reader to make. One of the Coordination Group members tells us, 'you have to do some imagination work, like it asks you from the outset "Walk with me up to the physical gates of this space, we will digitally open it together and explore it'" (Coordination Group member). The campaign invites the reader to put themselves in the place of detention, feeling what it would be like to be there; it asks you to imagine what it would be like if *you* were detained.

The Twitter campaign moves the reader through the spaces of detention itself, uncovering the realities that are often left unreported. In 2015, Unlocked began to run question-and-answer sessions on Twitter, with people who were at the time detained in IRCs the campaign visited. Members of the audience were encouraged to submit questions to Coordination Group members, who were then able to put them out to a person in detention, and 'tour' coordinators then posted the answers for the general public to read. Representing detainee voices in this way can be characterised as an important shift of control over the visibility of migrant populations from the Home Office to the Unlocked campaign, as the campaign facilitates the voices of detainees, permitting them to become present on social media. Whilst we acknowledge that control over the presence of these voices remains located with those outside of detention, rather than with detainees themselves, we believe that the process of co-creating a virtual presence for detainees is significant because it makes a novel form of collective action possible by engaging the general public in debates around the use of detention, and calling on them to act politically by participating in the campaign. In the same year, Unlocked also established a campaign that asked members of the general public to take a photo of something they would miss if they were detained without a trial or a time limit, and to post it on Twitter with the *hashtag* '#unlocked'. In a series of photographs that included children, partners, and grandparents, scenic landscapes and nature shots, photographs of people's homes, and moments of cooking or dancing with friends, all those who took part were asked to consider how they would feel if they were deprived of their liberty. In asking people to imagine how it would feel to be 'deprived of life' (Tyler, 2006: 189), just as those who are detained are deprived of their lives, the campaign aims to strengthen the emotional connections between those inside and outside detention. One Coordination Group member tells us: 'I think even that moment where you take the second to pose that question, you're in, to a certain extent. You're in the debate, you're in the thing' (Coordination Group member). By mobilising a virtual version of detention, the campaign hopes to create an immersive experience, inviting the reader to enter into the unknown world of the IRCs. Thus, detention becomes a *collective concern* rather than something anonymous, hidden, and imposed upon an unknown other (Ahmed, 2004); it allows friends, families, support workers, and communities to share their experiences of detention in order to speak with detainees, and on their behalf.

Moving beyond detention

Academic enquiry into the use of detention has conceptualised it as a liminal 'zone of abandonment' outside of sovereign law (Moran, 2012: 306), where basic rights are 'suspended' (Tyler, 2006: 189). Whilst the physical presence of those detained remains invisible to the public, the blogs provide a way of gaining insight into what happens within detention facilities, by mobilising what would be otherwise unheard experiences. Our research highlights various emotional dynamics within the blogs, but in particular we explore how individual narratives are constructed by detainees to challenge the physical immobility that they experience. The descriptions of the emotions felt by those within IRCs, such as fear, loneliness, uncertainty, and detachment, become a powerful way of engaging the reader by mobilising a non-physical version of their experiences. For example, in one blog post a detainee writes, 'I was isolated from everyone and everything I knew. With no way of speaking to anyone outside, I really felt forgotten' (Jamal, The Verne), and another detainee describes their experience of detention as 'mental torture' (anonymous, Morton Hall). Such descriptions of the emotional cost of detention reflect broader criticisms by non-governmental organisations, including a report by Medical Justice that described detention as a 'second torture' (Tsangarides, 2012), which draws attention to the fact that many people who are detained are survivors of torture or other atrocities in their countries of origin. The blogs offer new ways of mobilising how it feels to be detained, whilst also subverting the barriers that the Home Office imposes on those who are in detention, such as the ban on social media.

We have chosen two accounts by individuals who provide detailed accounts of their experiences in detention: Henry, who was detained in Haslar, and Abri who was detained in Yarl's Wood. We include an extract by each of these detainees to demonstrate how their personal narratives provide a challenge to mainstream views about the use of detention. These are presented verbatim, to preserve the original intentions of the authors.

Henry writes:

> My name is Henry, aged 53, and though of Nigerian and Sierra Leonean mixed parentage, I have spent the last 31 years of my life in the United Kingdom. I have a British wife, and two sons who were born here. I was arrested and detained since the beginning of Feb 2014 despite not having defaulted on anything for four years. I am considered a 'risk to society' by the Home Office because of a non-violent crime I committed, and served prison time for, 17 years ago (…) I am due to be deported to a country I have not visited for 31 years and in which I have no family or friends. The most painful thing about my incarceration, which will live in my memory forever, is that I am detained at a centre which is NINE hours away from my wife and six year old son. (…) Though I speak to my wife and son more than TWENTY times a day (and I do get frequent free calling credit from *Friends Without Borders* charity) it is never the same as seeing them, hugging them, crying with them, make jokes with them and sharing simple basic bonding; therefore each night I cry myself to sleep and pray to a God who seem to have deserted me at the age of 53 and hope for a better tomorrow. (Henry, Haslar IRC)

Posts like Henry's show how the campaign questions the human cost of detention for not only asylum seekers, victims of torture and trafficking, women and children, but also men who are detained having already served a criminal sentence.

What we find interesting about Unlocked is that it has not been reluctant to discuss the more complicated realities of detention; it explicitly attempts to counter the fact that coverage of the many types of people held in detention, including ex-national foreign prisoners, is often left unreported (Ohtani and Allsopp, 2014). The Coordination Group members hope that giving voice to a largely unheard population could generate a shift in attitudes, for example where someone could start to think 'ah, okay that guy had a really interesting thing to say, and he happens to be an ex-offender' (Coordination Group member), which is why developing a contextualised understanding of how detention is used that relates to individual circumstances may provide a serious political challenge to mainstream views.

Mobilising new forms of presence from individual identities, which would otherwise be concealed from the public domain, also exposes the psychological impact that comes from the material conditions of confinement and isolation. The emotions that individuals experience, such as the pain, anger, and frustration of being isolated, can form a powerful political message. This is further demonstrated by Henry, who writes in his second blog post, 'I've been in detention for eight months and I haven't set eyes on my family. It's more than difficult, it's traumatic', which contrasts sharply with the vision of detention framed as a non-punitive, administrative measure. The indefinite incarceration of people for administrative reasons raises broader issues of public concern about human rights and individual liberty. Unlocked aims to mobilise the voices of detainees in public space, in order to question the legitimacy of detention, and raises it as a matter for public concern, regardless of the individual gender, migration status or criminal record of someone who has been detained.

Abri's contribution to the blog further demonstrates how emotions become determined by the experience of detention, and how this is exacerbated by the fact that at present detention in the UK has no time limit. She writes:

> I was forced to run and flee for my life. The fact that I am a woman made me a victim of sexual assault in my home country. Now I'm locked up 24/7 indefinitely in Yarl's Wood, the future is uncertain and deep inside me the voice of hope is daily fading away. I'm separated from family and everything that is normal. I can't plan for tomorrow because I don't know where my tomorrow is (…) I'm one of the women detained in Yarl's Wood detention centre and these are not just my feelings, I'm voicing the feelings of many women like me who have been detained indefinitely for months or years. It is hard to put in words what it feels like. Every day something very valuable is taken from us – our freedom. You eat for comfort and sleep to escape, but struggle with both. Thoughts of failure, shame, guilt and defeat fill your mind. You have become so helpless you can't even choose your dinner as decisions are made for you daily. You lose touch with humanity and start to feel that human rights don't apply to you. (Abri, Yarl's Wood IRC)

For Abri, speaking out in this way is political action; she begins her second blog post by writing, 'today I would like to share some personal experiences and those of fellow detainees, with the hope that we can get people to stand in solidarity with us while we wait for justice'. This indicates how detainees may view the blogs as an important way of overcoming the isolation of the centres, by exposing the nature of their otherwise hidden incarceration, and hoping to gain political support for a resolution of their situation. The reach for a connection beyond the confines of detention to seek solidarity is explained by the Coordination Group:

> I think it connects those closest to the site with the world outside and it makes those membranes much more permeable, and that can only be a good thing. It feels like the arms stretching out over the fences, stronger, there's fluidity but there's also solidarity. (Coordination Group member)

Thus, where a lack of external scrutiny results in injustices for the detainees during their incarceration, the blogs themselves become an important vehicle for enabling those who are detained to voice their experiences outside the centres. This broadens the spaces of detention, making their boundaries more fluid, as the presence of detainees is reconstituted through online networks, de-stabilising the feelings of isolation that 'construct' the spaces as carceral (Moran, 2012: 306; see also Turner, 2016).

Conclusion

The Internet has proven to be crucial to Unlocked as it allows information about IRCs to be circulated, which would otherwise be unheard. We have uncovered three different types of mobility that arise from the campaign's use of online platforms. Firstly, we have demonstrated how the Internet can be used for political mobilisation. By revealing the experiences of detainees, the 'Unlocking Detention' campaign has been able to open up detention to public scrutiny, and has aided the dissemination of information about the widespread injustices detained populations face, enabling the issue to become visible in the public domain. Secondly, we have shown the different ways that presence may be mobilised through virtual space, using various creative tactics to overcome the technological restrictions placed on people within IRCs. This allows for the circulation of personal accounts of experiences of detention in novel ways that would otherwise be impossible or unsafe. Thirdly, we have explored how emotionally moving subjects may be used to mobilise political action; and how emotions can be felt together to co-create a space of presence. A primary achievement of Unlocked is the way it has been able to publicise the emotional violence of detention. The fear, loneliness, uncertainty, frustration, and pain that it produces run directly counter to government claims that detention is administrative and non-punitive in character. Rather, the emotional consequences of detention serve to remind various publics that 'detainees' are people who suffer as a result of expensive and bureaucratic border controls, not merely 'bodies' in carceral systems (Hall, 2012). With these points in mind, this chapter invites further exploration of the ways that online platforms may be able to assist in making audible other unheard voices of marginalised and excluded groups.

Notes

1 According to National Statistics published by the Home Office the number of people entering detention in the year ending June 2015 increased by 10 per cent to 32,053 (Home Office, 2015b).
2 This binary contrast itself is regrettable as it hides a whole host of other irregular migrants who get trapped in the detention estate.
3 A number of Freedom of Information (FOI) requests were submitted to clarify why websites are officially blocked by the Home Office. FOI release 32401 (Home Office, 2014) and FOI release 32251 (Home Office, 2015a) both state:

> The service provider at each [IRC] adheres to technical and functional specifications set by the Home Office which are designed to minimise any risk to the safety and security of the centre, to ensure that detainees are not exposed to offensive or inappropriate material and to protect the public from harm. Prohibited categories of sites include gambling, racist material, social networking, pornographic material and websites supporting and promoting acts of terrorism or containing extremist and radicalisation material. (Home Office, 2014; Home Office, 2015a)

4 There are number of blog posts on the Unlocking Detention website where both people who are detained and visitors talk about the experiences of friendship and mutual appreciation.
5 Standoff Films is an independent film production company with particular interest in UK detention. They not only provide an online platform for films and voices recorded at IRCs, but also make their own short films on detention.

References

Ahmed S (2004) *The Cultural Politics of Emotion*. Edinburgh: Edinburgh University Press.
Aouragh M and Alexander A (2011) The Arab spring| the Egyptian experience: Sense and nonsense of the Internet revolution. *International Journal of Communication* 5: 1344–1358.
Bail for Immigration Detainees (2009) *Out of Sight, Out of Mind: Experiences of Immigration Detention in the UK*. Available at: www.osservatoriomigranti.org/assets/files/BID%20-%20Out%20ofsight%20out%20of%20mind.pdf
Barnett C (2004) Media, democracy and representation: Disembodying the public. In: Barnett C and Murray L (eds) *Spaces of Democracy: Geographical Perspectives on Citizenship, Participation and Representation*. London: 185–206.
Bondi L (2003) Empathy and identification: Conceptual resources for feminist fieldwork. *ACME: An International E-Journal for Critical Geographies* 2(1): 64–76.
Bosworth M and Guild M (2008) Governing through migration control: Security and citizenship in Britain. *The British Journal of Criminology* 48(6): 703–719.
Büscher M and Urry J (2009) Mobile methods and the empirical. *European Journal of Social Theory* 12(1): 99–116.
Butler J (2011) Bodies in alliance and the politics of the street. *European Journal for Progressive Cultural Policies*. Available at: http://eipcp.net/transversal/1011/butler/en
Channel 4 (2015) *Yarl's Wood: Undercover in the Secretive Immigration Centre*. Available at: www.channel4.com/news/yarls-wood-immigration-removal-detention-centre-investigation
Coddington K and Mountz A (2014) Countering isolation with the use of technology: How asylum-seeking detainees on islands in the Indian Ocean use social media to transcend their confinement. *Journal of the Indian Ocean Region* 10(1): 97–112.
de Certeau M (1984) *The Practice of Everyday Life*. California, CA: University of California Press.

Detention Forum, The (2016) *Unlocking Detention*. Available at: http://detentionforum.org. uk/unlocked/

Eastmond M (2007) Stories as lived experience: Narratives in forced migration research. *Journal of Refugee Studies* 20(2): 248–264.

Fincham B, McGuiness and Murray L (2010) *Mobile Methodologies*. Basingstoke: Palgrave Macmillan.

Gill N (2009) Governmental mobility: The power effects of the movement of detained asylum seekers around Britain's detention estate. *Political Geography* 28(3): 186–196.

Gill N (2013) Mobility versus liberty? The punitive uses of movement within and outside carceral environments. In: Moran D, Gill N and Conlon D (eds) *Carceral Spaces: Mobility and Agency in Imprisonment and Migrant Detention*. Farnham: Ashgate, 19–36.

Gill N (2016) *Nothing Personal? Geographies of Governing and Activism in the British Asylum System*. Oxford: Wiley-Blackwell.

Gill N, Conlon D, Tyler I and Oeppen C (2014) The tactics of asylum and irregular migrant support groups: Disrupting bodily, technological, and neoliberal strategies of control. *Annals of the Association of American Geographers* 104(2): 373–381.

Griffiths M (2014a) *Immigration Detention in the media: Anarchy and Ambivalence*. Available at: www.opendemocracy.net/5050/melanie-griffiths/immigration-detentionin-media-anarchy-and-ambivalence

Griffiths M (2014b) *The Deserving Detainee? Available at: http://#unlocked.org.uk/blog/ the-deserving-detainee/.*

Guardian (2014) *Call For Inquiry into Death at Morton Hall Immigration Detention centre*. Available at: www.theguardian.com/uk-news/2014/sep/07/morton-hall-immigration-detention-centre-death-rubel-ahmed

Guardian (2015) *Home Office to Compensate Pregnant Asylum Seeker for Unlawful Detention*. Available at: www.theguardian.com/uk-news/2015/oct/06/home-office-to-compensate-pregnant-asylum-seeker-for-unlawful-detention

Hall A (2012) *Border Watch: Cultures of Immigration, Detention and Control*. London: Pluto Press.

Hannam K, Sheller M and Urry J (2006) Editorial: Mobilities, immobilities and moorings. *Mobilities* 1(1): 1–22.

Home Office (2014) *FOI Release: Internet Access Restrictions for Detainees in Immigration Removal Centres*. Available at: www.gov.uk/government/publications/ internet-access-restrictions-for-detainees-in-immigration-removal-centres/ internet-access-restrictions-for-detainees-in-immigration-removal-centres

Home Office (2015a) FOI Release: Internet Access Provided to Detainees in Immigration Removal Centres. Available at: www.gov.uk/government/publications/internet-access-provided-to-detainees-in-immigration-removal-centres/internet-access-provided-to-detainees-in-immigration-removal-centres

Home Office (2015b) *National Statistics: Detention*. Available at: www.gov.uk/government/ publications/immigration-statistics-april-to-june-2015/detention

Independent (2014) *Serco given Yarl's Wood Immigration Contract Despite 'Vast Failings'*. Available at: www.independent.co.uk/news/uk/politics/serco-given-yarl-s-wood-immigration-contract-despite-vast-failings-9880772.html

Jensen M and Jolly M (2014) (eds) *We Shall Bear Witness: Life Narratives and Human Rights*. Madison, WI: University of Wisconsin Press.

Kozinets RV (2015) *Netnography: Redefined* (2nd edition). London: Sage.

Madge C (2007) Developing a geographers' agenda for online research ethics. *Progress in Human Geography* 31(5): 654–674.

Martin LL and Mitchelson ML (2009) Geographies of detention and imprisonment: Interrogating spatial practices of confinement, discipline, law and state power. *Geography Compass* 3(1): 459–477.

Moran D (2012) "Doing time" in carceral space: TimeSpace and carceral geography. *Geografiska Annaler: Series B, Human Geography* 94(4): 305–316.

Mountz A (2011) Specters at the port of entry: Understanding state mobilities through ontology of exclusion. *Mobilities* 6(3): 317–333.

Mountz A, Coddington K, Catania, RT and Loyd JM (2012) Conceptualizing detention. Mobility, containment, bordering, and exclusion. *Progress in Human Geography* 37(4): 522–541.

Ohtani E and Allsop J (2014) *Migrant Lives in the UK: The Deprivation of Liberty.* Available at: www.opendemocracy.net/5050/eiri-ohtani-jennifer-allsopp/migrant-lives-in-uk-deprivation-of-liberty

Papacharissi Z (2002) The virtual sphere: The Internet as a public sphere. *New Media & Society* 4(1): 9–27.

Schaffer K and Smith S (2014) E-witnessing in the digital age. In: Jensen M and Jolly M (eds) *We Shall Bear Witness: Life Narratives and Human Rights.* Madison, WI: University of Wisconsin Press.

Silverman SJ (2012) 'Regrettable but necessary?' A historical and theoretical study of the rise of the UK immigration detention estate and its opposition. *Politics & Policy* 40(6): 1131–1157.

Simmel G, Frisby D and Featherstone M (1997) *Simmel on Culture.* London: Sage.

Smith K (2015) Stories told by, for, and about women refugees: Engendering resistance. *ACME: An International E-Journal for Critical Geographies* 14(2): 461–469.

Standoff Films (2015) *Standoff Films.* Available at: www.standoffilms.com/

Tsangarides N (2012) *'The Second Torture': The Immigration Detention of Torture Survivors.* London: Medical Justice.

Turner J (2016) *The Prison Boundary: Between Society and Carceral Space.* London: Palgrave Macmillan.

Tyler I (2006) 'Welcome to Britain': The cultural politics of asylum. *European Journal of Cultural Studies* 9(2): 185–202.

Tyler I (2010) Designed to fail: A biopolitics of British citizenship. *Citizenship Studies* 14(1): 61–74.

Unlocked (2015) *Unlocking Detention: A Virtual Tour of the UK's Immigration Detention Estate.* Available at: www.unlocked.org.uk

Urry J (2007) *Mobilities.* Cambridge: Polity Press.

Valentine G, Sporton D and Nielsen KB (2009) Identities and belonging: A study of Somali refugee and asylum seekers living in the UK and Denmark. *Environment and Planning D: Society and Space* 27(2): 234–250.

Wilsher D (2008) The administrative detention of non-nationals pursuant to immigration control: International and constitutional law perspectives. *International and Comparative Law Quarterly* 53(1): 897–934.

Wooley A (2014) *Contemporary Asylum Narratives: Representing Refugees in the Twenty-First Century.* London: Palgrave Macmillan.

11 Mobile authority

Prosecutorial spaces in the Parisian *banlieue*

Joaquín Villanueva

Introduction

On April 23, 2008, four individuals were placed under police custody after more than 350 officers were deployed in the Beaudottes social housing project, 'a sensitive neighbourhood of Sevran', a commune in the Department of Seine-Saint-Denis, northeast of Paris, France (*Libération*, 2008: n.p.; See Figure 11.1). The French Intelligence Service had notified criminal justice personnel that the Beaudottes was in an 'explosive situation', with drug dealers imposing their violent will against inhabitants and, most importantly, authorities unable to penetrate, police and regulate the everyday spaces of the housing complex (350 policiers, 2008). Thus, on this spring day of 2008 various jurisdictions from Paris, Seine-Saint-Denis, and Sevran joined forces in an attempt to reassert the authority of the government in this so-called 'priority neighbourhood' (Mandraud, 2008).

Under the gaze of the police prefect and the chief prosecutor, the officers conducted the search and surrounded 'eight building halls that had been identified as regular drug selling spots and a dozen empty apartments that were targeted as storage rooms for narcotics' (*Libération*, 2008: n.p.). The Beaudottes had been under investigation by a Local Group for the Treatment of Delinquency (*Groupes Locaux du Traitement de la Délinquance*, GLTD), a public partnership presided by the chief prosecutor which had been operational since September of 2007. After months of close surveillance, officials seized only small amounts of marijuana, a handgun, and some stolen items. Despite the little evidence collected, the GLTD justified the operation on the grounds that, as a magistrate expressed, 'we needed to exploit the information that was at our disposal' (Mandraud, 2008: n.p.). Unsaid was the fact that, as I claim, the GLTD at Beaudottes was an exercise in search of recognition for state authority, an activity that required prosecutorial mobility.

In this chapter I argue that prosecutorial mobility – the ability of the chief prosecutor to not only be physically present in this vast police operation, but her/his ability to move flexibly across the urban landscape with the intention of mobilising local and regional resources destined to contain or detain individuals located in 'priority neighbourhoods' – has become an important tactic for the exercise of judicial authority in the Parisian *banlieue*. I draw on Hannah Arendt (1961) to conceptualise authority, a form of power that requires recognition for the person or office

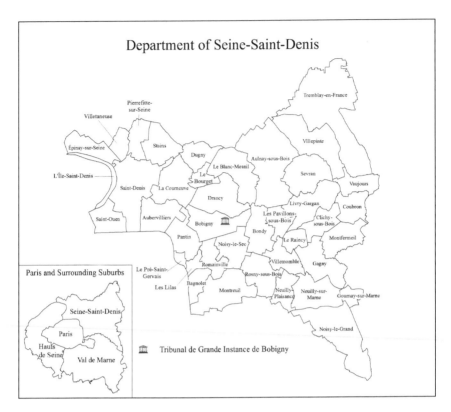

Figure 11.1 Communes of Seine-Saint-Denis.

(Map prepared by David Brown.)

claiming it in order to be effective. Recognition of authority is more likely, as John Allen (2003) has expressed, when in close proximity to those one seeks acknowledgement and respect. Prosecutorial mobility is thus a means to approximate subjects and institutions to, among other ends, (re)establish authoritative relationships.

Prosecutorial mobility in the Parisian *banlieue* has been facilitated by the GLTD, a public partnership presided by the chief prosecutor that operates for a short period of time (6 to 24 months) in a specified area (park, neighbourhood, school surroundings, etc.). The GLTD is a space of information and resource exchange that brings together, under the guidance of the prosecutor's office (PO), local partners such as the police, elected officials, school administrators, and, in some cases, private business owners with the ultimate goals of (1) re-establishing the authority of the law, and (2) creating 'tolerable living conditions' in the areas it operates (Body-Gendrot, 2001: 57) These goals are pursued, first, by a repressive approach to crime and violence which seeks to arrest and swiftly prosecute individuals visibly breaking criminal laws and, second, a preventative phase where neighbours are encouraged to denounce any future wrongdoings to the authorities and to trust state institutions. Once partners of the GLTD determine that they have accomplished these tasks, the PO moves its operations elsewhere. The GLTD is, in short, a judiciary in motion (*Le Monde*, 1998).

The chapter calls for greater attention to be paid to prosecutorial spaces – district courts, community courts, juvenile courts, and judges' and prosecutors' offices, among many more underexplored locations – and their mobility practices. Prosecutorial mobility, I demonstrate, has expanded the geographies of judicial authority and has brought a larger number of individuals and locations *closer* to detention and incarceration. In the next section I begin to place prosecutorial spaces within the context of the carceral geographies literature and suggest we conceive them as part of the complex 'transcarceral spaces' that together constitute our carceral societies (Allspach, 2010; Brown, 2014). I then conceptualise prosecutorial mobility around Arendt's definition of authority. I proceed to introduce the case of an inter-communal GLTD in Montreuil and Bagnolet (Seine-Saint-Denis) that was in operation in 2007. I show that by vesting its judicial authority on local actors, the PO seeks to narrow the spatio-temporal gap between adjudicators and the adjudicated. I conclude by suggesting that due to its inability to be 'spatio-temporally fix', prosecutorial mobility, through the GLTDs, cannot effectively establish authoritative relationships across time and space.

Prosecutorial spaces in carceral geographies

Prosecutorial spaces form an integral part of the complex socio-spatial 'processes of detention and confinement' (Martin and Mitchelson, 2009: 459). Nevertheless, with some notable exceptions (Brown, 2014), carceral geographers have surprisingly neglected prosecutorial spaces as key sites for the practices of surveillance, disciplining, detention, and confinement. When taken into consideration, prosecutorial spaces have been mostly understood as sites where the decisions pronounced constrain the mobility of individuals elsewhere (Beckett and Herbert, 2010; Cresswell, 2006). Such investigations assume prosecutorial spaces as 'fixed' in space, without considering the mobility of judges, prosecutors, and their offices. I suggest we conceive prosecutorial spaces through the lens of 'carceral mobilities' to help grasp their full spatial reach.

Research on the spatialities of carceral spaces has garnered significant attention by critical geographers in recent years (see Chapter 1, this volume). Allspach (2010), in particular, introduced the notion of 'transcarceral spaces' to help appraise the complex geographies of carcerality. Drawing on in-depth interviews with federally sentenced women in Canada, Allspach studied the pre-prison, prison, and post-prison experiences and everyday geographies of women which had been subject to racialised, gendered, classed, and criminalised practices along this long trajectory of social control. She concluded that 'the practices of social control after women's release from federal prisons in Canada produce continuities of confinement beyond prison walls on multiple levels' (Allspach, 2010: 721). Similarly, Dominique Moran explored the embodiment of imprisonment by ex-offenders in Russia and noted that 'transcarceral spaces exist alongside and perhaps also in combination with an embodied sense of the "carceral" which is similarly mobile beyond the prison wall through the corporeality of released prisoners' (2014: 37).

Brown made similar remarks but went further in calling for a more 'geographically expansive consideration of incarceration' (2014: 377, see also this volume). In her investigation of juvenile detention in California, Brown compellingly argued that carceral geographers 'must also consider how the multiplicity of carceral spaces ultimately work together to create the carceral society that is so common to modern day descriptions' (2014: 386). Broken windows and community policing, therapeutic courts, and residential placements are some of the 'transcarceral spaces', programs and institutional arrangements that have worked to extend the reach and scope of social control in North America and Western Europe. The vastness of social control and punishment practices which have brought the experience of incarceration closer to the lives of the poor and communities of colour have been extensively explored by scholarship on the emergence of the carceral state – a set of institutional configurations and actors that prioritise punishment, containment, detention, and/or incarceration for treating poverty and marginality (Beckett and Murakawa, 2012; Peck, 2003; Wacquant, 2009). Moreover, the expansion of surveillance, control, and the threat of incarceration through court fees, probation and parole, substance-abuse treatment, and practices of 'banishment' all point to the creative and extensive reach of the carceral state beyond prison walls (Beckett and Herbert, 2010; Belina, 2007; Goffman, 2014).

In recent years, critical scholars have tended to look beyond the prison to comprehend the complex socio-spatial processes of detention and incarceration. Under these geographical arrangements, prosecutorial spaces, and those that embody prosecution (judges and prosecutors, in particular), constitute an important link and an enabling force in the 'continuities of confinement' – whether before or after prison time. As such, we can conceive prosecutorial spaces as 'a single node in larger practices of social control' (Brown, 2014: 385). And yet, as I argue, prosecutorial spaces can be a *moving* node in the vast networks of social control that constitute the 'republican penal state' (Dikeç, 2006). Distinct from its American counterpart, the 'republican penal state' is characterised less by the use of the prison as a strategy of crime-control, poverty containment, and racial differentiation. Instead, punishment '*à la française*' is mostly 'effected by means of the police and courts' (Wacquant, 2001: 407). Punishment *à la française* has also entailed the increased mobility of the courts *into* 'priority neighbourhoods'. This movement, as shown below, has expanded the geographies of state authority and brought a greater number of subjects closer to incarceration.

Prosecutorial mobility and the extension of judicial authority

In her investigations into the detention, prosecution, and media outrage over hundreds of Fujianese refugee claimants that came by boat to Canada in 1999, Alison Mountz (2010) showed how the Canadian state temporarily reconfigured the legal landscape in order to keep claimants outside sovereign territory. By placing refugee claimants in a remote detention facility in British Columbia, authorities ensured that asylum-seekers did not get adequate *access* to the legal representation they were entitle to. On the other hand, the Immigration and Refugee Board of Canada (IRB) had no problem finding adjudicators to lead the tribunals that eventually determined the refugee status of hundreds of underrepresented claimants. Hearings for these claimants 'were held in provisional tribunals established within the prison on Prince George and adjudicated by officers of the IRB who were *flown in*' (Mountz, 2010: 108; emphasis added). Prosecutorial mobility was a strategy adopted by the Canadian state to effectively accelerate the determination process of hundreds of individuals who were eventually deported to China. Prosecutorial mobility, as I argue, can be a means for reasserting state (and judicial) authority *with* subjects and places that otherwise seek to elude such power.

For Hannah Arendt (1961), authority is but one form of associational power because it is 'held *among* people, not over them' (Allen, 2003: 58; emphasis in original). Authority, moreover, 'can be vested in persons ... or it can be vested in offices' and, once vested, authority demands 'unquestioning recognition by those who are asked to obey' (Arendt, 1970: 45). Thus, the IRB office demanded recognition from the refugee claimants who forcibly recognised its claim to authority. 'Once claimed,' John Allen reminds us (2003: 6), authority has to 'justify itself in the eyes of those around them' because compliance, respect, and the recognition of authority are 'always conditional'.

In both Canada and France, authority claims were made in person, in close proximity of those the IRB and the GLTD sought recognition from – the Fujianese claimants and the Beaudottes residents, respectively. Proximity is a crucial element for authority's effectiveness. Allen (2003: 148) noted that 'the authority-recognition relationship is less effective – not ineffective – as it is more complexly mediated, more distant'. Furthermore, 'authority's constant need for recognition implies that the more direct the presence, the more intense the impact' (Allen, 2003: 149). How can the judiciary, I ask, successfully mediate 'authority relations when confronted with a diverse and dispersed civic population' such as in the Parisian urban region (Allen, 2003: 149)? The answer is prosecutorial mobility, a strategy that has enabled the French judicial system to move closer to the spaces and individuals that it wishes to strengthen authority relations with. As explained below, GLTDs in 'priority neighbourhoods' are mobilised in and through authoritative relationships *with* local actors who then act in the city on behalf of the PO. The GLTDs short duration coupled with the PO's indirect presence, I argue, prevent it from being 'spatio-temporally fix' which thus reduces the effectiveness of judicial authority across time and space.

Mobile authority: The *Groupes Locaux du Traitement de la Délinquance* (GLTD)

As the unofficial story goes, at the end of 1992 the mayor of Stains, a small commune in Seine-Saint-Denis received a letter from the main supermarket chain. Carrefour gave the municipality an ultimatum: control the repeated thefts and aggressions caused by the 'quotidian' presence of youngsters in their premises or they leave. The commune took matters seriously given the lack of viable economic alternatives. Yet the municipality was unable to provide the necessary means of security to Carrefour, so it turned to the prosecutor's office at the district court of Bobigny (POB) to evaluate its options. Thereafter, the POB, the police prefect, the municipality, and the directors of Carrefour became partners under the first Local Group for the Treatment of Delinquency in 1994. The GLTD associates the competences, resources, and know-how of local actors and state institutions to collectively 'restore social peace in neighbourhoods affected by an endemic violence' (Prieur, 1998a: n.p.).

During monthly meetings *at the tribunal*, partners devise strategies to reduce the 'feeling of insecurity' and 'restore' social relationships in a well-defined area, such as a housing estate, a school, a park, or a street (Donzelot and Wyvekens, 2004). To accomplish these objectives, the GLTD combines repressive and preventative measures over a specific duration of time that could well span from six to 24 months. The first phase of the GLTD consists in identifying the objectives of the operation. These could widely range, but examples include stemming the sources of violent crime, controlling the underground economy, and treatment of school absenteeism (Action Publique et Sécurité, 2001). The second phase consists in the deployment of police officers in the area of intervention. At this point arrests are made and detainees are referred to the district court where they are swiftly prosecuted. At the beginning, explains Pierre Debue, former director of public security at Seine-Saint-Denis, it is imperative to 'show troublemakers that they are not "the kings of the neighbourhood"' (Ceaux, 1998: n.p.). During the initial phases, the GLTD actively searches for recognition of judicial authority in places where the prosecutor's power seems threatened, delegitimised or, worse, unrecognised. The swift removal *and* prosecution of lawbreakers demonstrates that the authority of the 'republican law' has been re-established and that the third phase is under way.

Following the repressive period, partners of the GLTD organise preventative measures aimed at restoring social relations in the neighbourhood. Measures range from investments in the physical façade of buildings, prevention campaigns in schools, and sessions on parental authority intended to encourage inhabitants to denounce any wrongdoing to the police and the courts. A manager of an HLM (*Habitation à Loyer Modéré*, Rent-Controlled Housing) at Stains observed that 'before [the GLTD], people turned to the HLM when a problem arrived. Now, they go to the police and the judiciary because they finally have the impression that they are taken into account' (Prieur, 1998b: n.p.). The recognition of the police and courts by inhabitants as trustworthy institutions is a clear sign that the conditions for associational power relations have been met.

Once judicial authority relations have been seemingly established in the area, the prosecutor can terminate the GLTD. Pierre Moreau, the chief prosecutor who launched the first GLTD in 1994, noted that the role of this partnership is to 'reestablish social cohesion when it seems truly threatened in a neighbourhood, by identifying and responding to the sources of insecurity. Once the major problems have been solved, we reinvest [our resources] in another place' (Prieur, 1998a: n.p.).

Former Interior Minister, Jean-Pierre Chevènement, clarified that GLTDs cannot commit too long in one neighbourhood because this could foster the 'transfer of delinquency to [other] neglected areas' (Chevènement, 1998: 1717). GLTDs constitute, added Chevènement, 'surgical solutions' to the 'hottest' neighbourhoods with the ultimate goal of 'giving back the population confidence [in the capacity] of public authorities' to solve their problems (Chevènement, 1998: 1717). In other words, GLTDs are expected to fix the security problems of 'priority neighbourhoods' without necessarily being temporally or spatially 'fixed'.

During the 1990s, GLTDs were lauded by local elected officials, the courts, the police, the media, and critical scholars for their effectiveness in treating crime and violence (Body-Gendrot, 2001). By the turn of the century, GLTDs faded from view despite the enactment of the circular of May 9, 2001 which set the administrative parameters for creating and operating a GLTD (Action Publique, 2001). The post-2005 period saw a resurgence of GLTDs in response to the 'urban riots' that swept the French landscape (Matthew Moran, 2011). With the creation in 2012 of the *Zones de Sécurité Prioritaires* (ZSP), designated areas provided with additional resources to fight insecurity, the GLTDs have found a new institutional home. Still presided by the chief prosecutor, GLTDs have been increasingly operated within the context of ZSPs, drawing on the vaster resources that this new geography of security facilitates (see Figure 11.2).

The present research is situated in the post-2005 context. I spent ten months (2007–2008) in Seine-Saint-Denis and investigated judicial responses to crime and violence in various communes. My knowledge of GLTDs was mainly acquired through interviews with judicial personnel and local officials directly involved with actually existing GLTDs at the time. Below, I present the case of Montreuil-Bagnolet, the first inter-communal GLTD, to demonstrate the complexly mediated power relations that seek to narrow the distance between adjudicators and the adjudicated.

Montreuil-Bagnolet GLTD

La Noue is a neighbourhood located in Montreuil that stretches to the adjacent commune of Bagnolet. In 2001, the community police at La Noue had disappeared 'leaving a big void' in the neighbourhood (Maesano, quoted in *Montreuil Dépêche Hebdo*, 2007: 13). The police prefecture of Montreuil had lost more than forty officers since 2001, a reduction which meant 'less human presence in the terrain and, by consequence, less prevention' (Maesano, quoted in *Montreuil Dépêche Hebdo*, 2007: 13). Over the years, less police presence led to an increased number of young people hanging out in the common areas, which contributed to the feelings of insecurity that had plagued this neighbourhood ever since. As the director of security at Montreuil described it:

> We had a very strong tension between youngsters, rival gangs that confronted regularly, armed with iron sticks, Molotov cocktails, etc. in short it was very, very violent. The neighbourhood of La Noue extends to the city of Bagnolet,

Geographies of Security and Marginalization in Seine-Saint-Denis

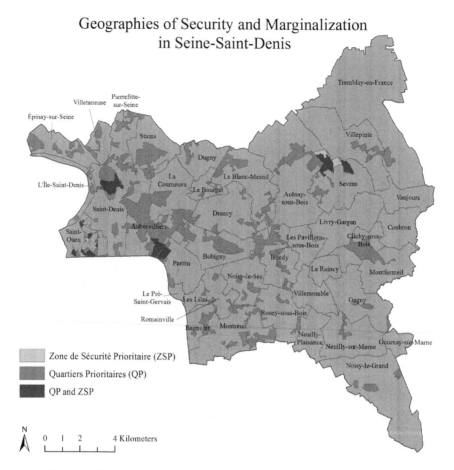

Figure 11.2 Priority neighbourhoods (QP) are government-designated areas with a large concentration of poverty. Priority security zones (ZSP) were created in 2012 to identify areas that needed additional resources to fight delinquency and insecurity. In Seine-Saint-Denis the geographies of poverty and insecurity overlap.

(Source of data: Ministère de la Ville and Ministère de l'Intérieur. Map prepared by David Brown.)

and that entertained a certain number of rivalries in a school at Montreuil that receives at the same time the youngsters from Bagnolet and Montreuil, so these rivalries went even inside the school premises, and in the vicinity of the school. (Interview, security official, Montreuil, 2008)

Having campaigned for a GLTD since 2005, the prosecutor's office of Bobigny (POB) finally gave green light to the project in 2006 and the first meeting of the group took place in February, 2007. The group included the chief prosecutor, the police prefects of Bagnolet and Montreuil, the director of the school, representatives of the HLM of La Noue, representatives of the mayors' office of Bagnolet and Montreuil, and local associations of the neighbourhood (Interview, Maesano, 2008). The GLTD lasted ten months and it was intended to respond to the increasing 'gang' violence and squatters in the area.

Vested authority

Former adjunct of security at Montreuil, Antoine Maesano, noted that during the GLTD all 'the meetings took place in the tribunal, because that gave an image' (Interview, Maesano, 2008). In the meetings, partners were able to 'follow-up on the youths at the school' (Interview, security official, Montreuil, 2008). They identified two profiles, 'on the one hand, we had kids in danger, in great difficulties, and, on the other hand, we had delinquent kids'. For partners, it was imperative to prevent the 'temporary exclusion from the school', because 'if one student is excluded they find themselves free outside and then he can perform delinquent acts in the neighbourhood' (Interview, security official, Montreuil, 2008). In other words, it was very clear from the outset that a central partner for this particular GLTD was the National Education – the institution in charge of public schools in France. Working alongside school administrators and teachers was imperative in order to monitor, surveil, and police young people *inside* the school. Thus, the POB attempted to vest its authority on school personnel as a means to bring these youngsters closer to the purview of the criminal justice system.

In this and other GLTDs, it is common practice for the PO to ask teachers to 'report' back to them (Donzelot and Wyvekens, 2004: 50). Reporting to the court entails giving up names of students and their behaviour inside the school premises and the classroom. The prosecutor, to ease any doubts the school might have with sharing evidence, often assures educators that not all information given will be subject to a 'penal' response. Information about school children, explains a prosecutor, is often use 'to recall young people what life in society, the law, is, so that [schoolteachers] don't always find themselves in the front line' (quoted in Donzelot and Wyvekens, 2004: 51). As Donzelot and Wyvekens argued in their research of GLTDs in the 1990s, schoolteachers who end up cooperating with the judiciary often view this partnership as an instance for 'reinforcing their authority' (2004: 52). This reinforced authority is particularly beneficial when dealing with 'trouble' students who can now be subject to disciplinary techniques beyond the seemingly ineffective measures of the school. Teachers vested with judicial authority do not enjoy the power of prosecution or adjudication. However, they could easily refer particular students to the POB, where their actions could be potentially subject to an array of penal responses, including incarceration.

Nevertheless, many teachers find it particularly uncomfortable to play the role of 'informants' of the POB. As a security official explained:

At the level of the National Education it is true there was some distance, they really didn't have the desire to denounce the youths with which we had difficulties because they didn't want to deal with all that had to do with the law, prevention of delinquency, etc. They were sort of reluctant, so we needed to show them that we were there to help them, and in fact they realized that the names that were being thrown out by the social landlords or the police [coincided] with the names of youths that they knew were problematic or who were suffering from absenteeism. Well, it's true that there was this reluctance from the part of the National Education, but they came regularly to the meetings. [From the outset] they knew that the interest of the GLTD is to provide an immediate response to crime.

The GLTD is more than a space of resource and information exchange. It is an associational power structure that allows the POB to move and extend the carceral logic *beyond* the limited confines of the district court. By virtue of it being able to vest its authority on local actors, the latter can act in the city on the POB's behalf. Despite the reluctance to fully assume the role of 'informants', educators use their added authority to further the surveillance, monitoring and disciplining of students on the fence between school and street life. Educators vested with judicial authority, in short, help mobilise the carceral geographies of prosecution *inside* the school, thus transforming the latter into a transcarceral space, albeit temporarily.

Expansive carceral geographies

In 2007, at the outset of the GLTD at La Noue, a police operation was launched where 'many were detained' which brought 'the calm back to the project' (Interview, security official, Montreuil, 2008). However, as many local officials lamented, the post-prison experience was no longer the responsibility of the prosecutor's office which by the time of my interviews had seized operations at Montreuil and moved its resources elsewhere. The GLTD is very effective at narrowing the spatio-temporal gap between arrest and prosecution since both the police and the courts work closely during that time to ensure the continuity of detention-incarceration. The sudden departure of the GLTD after ten months of operations meant that it was up to the municipality and local associations to manage the reintegration of ex-prisoners in the already fragile social networks of the neighbourhood.

'Today', said a security official at Montreuil, 'we are working with prison exit'. Those who had 'done prison-time', the official continued, 'often came back to the neighbourhood [bragging] about their time and claiming to be the strongest. We fear that younger brothers and youngsters will imitate them'. The municipality deployed social workers in the area to prevent such scenarios, yet, as the official admitted, 'the gathering in the building halls has not changed at all' (Interview, security official, Montreuil, 2008). The momentary fixity of GLTDs and its inability to 'fix' problems long term frustrated many local officials who believed to have found a 'miraculous' solution to crime and insecurity only to realise that it was only temporary. Moreover, during the GLTD, local officials bemoaned the fact that the incarceration of youths was not followed-up with measures of reintegration and/or restoration. In their eyes, containment for containment's sake was the POB's policy. 'When there was a judgement pronounced for a kid', said a security official, 'there was no real follow-up after it. It's not just about repression; the kids needed help to navigate [the administrative landscape]'. As a result, many of those arrested during the GLTD ended up engaging in the same activities that temporarily put them behind bars – squatting common areas, drug consumption, and physical altercations with rival 'gangs'.

Montreuil is perhaps one of the most 'dynamic' municipalities in managing 'alternative measures to the prison' (Interview, security official, Montreuil, 2008). The city funds, along with the Ministry of Justice, over twenty community service (*travaux d'intérêt généraux*, TIG) and 'penal reparation' programs designed to better integrate first-time offenders back into the city through non-paid work and educational measures (Locqueneaux, 2011: 6). These two examples are part of the so-called alternative measures to the prison, an array of judicial decisions, programs, and responses to crime that seek to provide *a* judicial response to offenses, however minor, while avoiding the financially, temporally, and geographically costly solution of incarceration (Interview, Judge Bobigny, 2008). Alternative measures to the prison have expanded the transcarceral spaces at the municipal level. They have become a veritable 'tool' for the extension of judicial authority through the monitoring of a larger number of individuals whose actions would have otherwise not been subject to a judicial response (see Figure 11.3). The multiple sites scattered across Montreuil in charge of following-up first-time

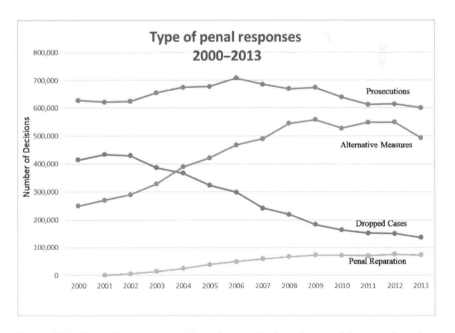

Figure 11.3 Alternative measures and penal reparation have increased the rate of penal response. In 2000 67.9 per cent of prosecutable cases received a penal response compared to 89.6 per cent in 2013.

(Source of data: Ministère de la Justice. Chart prepared by author.)

offenders and managing their reparation measures play a crucial function in the 'fight against the feeling of impunity'. 'It's not that we are against repression or against the prison', assures a security official, 'but it is necessary that before their release [from the system] there is a constructive approach' (Interview, security official, Montreuil, 2008).

Too much mobility, not enough presence

For local officials, the increasing involvement of the judiciary in local criminal affairs is deemed necessary, on the one hand, and insufficient on the other. Antoine Maesano expressed this frustration when evaluating the GLTD after its departure: 'We had one youngster that had many problems, we had an open file on him and it was clear that he needed all the available services to solve his problems … [Yet] we couldn't do anything because the judiciary didn't have the means' (Interview, Maesano, 2008). Mr Maesano further added that 'the judiciary is engaged, they provided support during the GLTD, but I don't think they have the means to be completely engaged'. Local officials recognised that the POB 'did not possess the human capacity, they lacked personnel'. 'We met once or twice a month,' Antoine Maesano frustratingly described it to me, 'but at times we couldn't get a hold of them because they didn't have time' (Interview, Maesano, 2008).

Local officials regard the judiciary as a distant institution that despite its limited resources desires to control every facet of the conflict-resolution process. In particular, the judiciary is viewed with scepticism when it comes to sharing information. Municipalities increasingly asked the POB to inform them about follow-ups on particular cases since many of the services at the disposal of that person are controlled and delivered by the municipality. Expressed a security official at Montreuil:

> I remember during the GLTD at La Noue we had the case of a woman that was victim of serious domestic violence by her husband. We wrote letters, we signalled it through the GLTD, etc. I mean, I am not saying that nothing was happening behind closed doors, but we had no reply on the monitoring of this person, or on how was this person being helped … It is a real pity, we know that the violence persists and we can't remedy it. (Interview, security official, Montreuil, 2008)

For Antoine Maesano, the GLTD is an excellent mechanism to 'manage the crisis'. 'I am happy with the GLTD', he continued, 'it worked for managing the crisis, but I think that the GLTD must be implemented in the long term' (Interview, Maesano, 2008). The momentary spatial fixity of the GLTD, furthermore, is what most commentators regret about this partnership approach to crime. A security official at Montreuil commented about the post-GLTD period at La Noue:

> Well it is true that the municipality is 'happy'. However, we had limited success at La Noue because the GLTD finished one day and still the problems

continued. The squatters in the building hall are very problematic. There was a strong consumption of alcohol in the street, the forces of order were assaulted, I mean, these were real problems. I think the GLTD was a failure because we weren't able to increase [the human presence in the neighbourhood]. In short, we tried to do a lot of stuff but as of today I think we have failed, and we haven't found a miraculous solution. The GLTD stopped one day and it's hard to find solutions [to crime] in [10] months. (Interview, security official, Montreuil, 2008)

Despite the increasing mobility of the courts into 'priority neighbourhoods', the POB is unable to consistently maintain its presence across time and space. The limited spatio-temporal fixity of GLTDs makes judicial authority 'less effective – not ineffective' in its ability to 'fix' the problems of crime, violence, and insecurity at the municipal level (Allen, 2003: 149). Critiques to the judiciary and its limited human capacity are in fact a sign that its effectiveness is in question. Yet, this questioning does not mean that judicial authority is unrecognised or disrespected by local officials. When asked if the judiciary is a legitimate partner in the GLTD, Antoine Maesano laughed at my question and pointed to the obviousness I had just missed: 'of course the judiciary is legitimate, it is *the* judiciary' (Interview, Maesano, 2008).

Conclusion

There were three running arguments in this chapter. First, I maintained that carceral and mobility scholars had not paid sufficient attention to prosecutorial spaces and, even less, to its practices of mobility. Second, this chapter argued, through the case study of the GLTD, that in France the prosecutor's office has the ability to move across the urban landscape in an effort to bring a greater number of individuals under the purview of the criminal justice system. With the passing of a GLTD, arrests are quickly followed by prosecution which puts individuals in a long trajectory of social control that for some include incarceration and further monitoring at the municipal level after prison time. Finally, this chapter discussed the geographies of authority of prosecutorial mobility. Authority was both a means *and* end of the POB. By vesting its authority on local actors, such as school personnel, the POB was able to narrow the spatio-temporal gap between adjudicators and the adjudicated. The POB acted in the city in and through local actors with the ultimate goal of re-establishing authoritative relationships with other local institutions and inhabitants that had grown sceptical of the POB.

Prosecutorial mobility, however, was less effective in extending the authoritative geographies it originally intended. Being that respect and recognition of authority are always conditional, authority requires a more direct presence to maintain that fragile association. Effective authority requires a 'spatio-temporal fix'. The rapid mobility of GLTDs, though, is not conducive to a lasting presence. For even if the security problems reoccur after its departure, there is no assurance the POB will return to that location to 'fix' them in the future. Moreover, their strategy of acting in the city *indirectly* through other actors reduces its chances of being recognised and respected by residents who feel insecure in their neighbourhood. The GLTD is effective, however, in bringing a greater number of individuals under the purview of the criminal justice system by virtue of it being a *moving* node with the capacity of hauling those near it into the networks of social control of the republican penal state. For that reason alone, carceral and mobility scholars must study prosecutorial spaces more attentively.

References

Action Publique et Sécurité (2001) Circulaire CRIM 2001-04 E/09-05-2001. *Bulletin du Ministère de la Justice*, 82, 1 April – 30 June.

Allen J (2003) *Lost Geographies of Power*. Oxford: Blackwell Publishing.

Allspach A (2010) Landscape of (neo-) liberal control: The transcarceral spaces of federally sentenced women in Canada. *Gender, Place, and Culture* 17(6): 705–723.

Arendt H (1961) *Between Past and Future: Six Exercises in Political Thought*. New York, NY: The Viking Press.

Arendt H (1970) *On Violence*. New York, NY: Harvest Book, Harcourt Brace and Company.

Beckett K and Herbert S (2010) *Banished: The New Social Control in Urban America*. Oxford: Oxford University Press.

Beckett K and Murakawa N (2012) Mapping the shadow carceral state: Toward an institutionally capacious approach to punishment. *Theoretical Criminology* 16(2): 221–244.

Belina B (2007) From disciplining to dislocation: Area bans in recent urban policing in Germany. *European Urban and Regional Studies* 14(4): 321–334.

Body-Gendrot S (2001) *Villes: La Fin de la Violence?* Paris: Presses de Sciences Po.

Brown E (2014) Expanding carceral geographies: Challenging mass incarceration and creating a 'community orientation' towards juvenile delinquency. *Geographica Helvetica* 69: 377–388.

Ceaux P (1998) A la cité des 4000, la police en butte aux 'rois des quartiers' de La Courneuve. *Le Monde*, 6 March. Available at: www.lemonde.fr

Chevènement J-P (1998) Réponse du ministère: Intérieur. *Journal Officiel, Sénat*, May 28.

Cresswell T (2006) The right to mobility: The production of mobility in the courtroom. *Antipode* 38: 735–754.

Dikeç M (2006) Two decades of French urban policy: From social development of neighbourhood to the Republican penal state. *Antipode* 38(1): 59–81.

Donzelot J and Wyvekens A (2004) *La Magistrature Sociale: Enquêtes sur les Politiques Locales de Sécurité*. Paris: La Documentation Française.

Goffman A (2014) *On the Run: Fugitive Life in an American City*. Chicago, IL: The University of Chicago Press.

Le Monde (1998) Quand la justice se mobilise. *Le Monde*, 25 April. Available at: www.lemonde.fr

Libération (2008) 350 policiers investissent une cité, 4 dealers présumés interpellés. Libération, 24 April. Available at: www.liberation.fr

Locqueneaux A (2011) L'Intérêt collectif de la réparation pénale. *Tous Montreuil* 48, 18–31 January, p. 6.

Mandraud I (2008) L'opération anti-dealers de la police à Sevran affiche un maigre bilan. *Le Monde*, 25 April. Available at: www.lemonde.fr

Martin LL and Mitchelson ML (2009) Geographies of detention and imprisonment: Interrogating spatial practices of confinement, discipline, law, and state power. *Geography Compass* 3(1): 459–477.

Montreuil Dépêche Hebdo (2007) GLTD: pour lutter contre la délinquance et s'attaquer aux racines des problèmes. *Montreuil Dépêche Hebdo*, 422, 31 January – 6 February, p. 13.

Moran D (2014) Leaving behind the 'total institution'? Teeth, transcarceral spaces and (re) inscription of the formerly incarcerated. *Gender, Place, and Culture* 21(1): 31–51.

Moran M (2011) Opposing exclusions: The political significance of the riots in French suburbs (2005–2007). *Modern and Contemporary France* 19(3): 297–312.

Mountz A (2010) *Seeking Asylum: Human Smuggling and Bureaucracy at the Border*. Minneapolis, MN: University of Minnesota Press.

Peck J (2003) Geography and public policy: Mapping the penal state. *Progress in Human Geography* 27(2): 222–232.

Prieur C (1998a) En Seine-Saint-Denis, une tentative de restauration de la paix sociale. *Le Monde*, 6 March. Available at: www.lemonde.fr

Prieur C (1998b) La vie trop brève du GLTD de la cité du Clos Saint-Lazare de Stains. *Le Monde*, 6 March. Available at: www.lemonde.fr

Wacquant L (2001) The penalisation of poverty and the rise of neo-liberalism. *European Journal on Criminal Policy and Research* 9(4): 401–412.

Wacquant L (2009) *Punishing the Poor: The Neoliberal Government of Social Insecurity*. Durham, NC: Duke University Press.

12 The other side of mobilities

Aboriginal containment in Australia from rail to jail, past and present

Katie Maher

Introduction

This chapter interrogates how shifts in the mobility and containment of Aboriginal populations at the turn of the twentieth century, are imprinted on present-day regimes of carceral control. Focusing on mobility and containment through a case study of the relationship between railways and incarceration, in the contexts of colonial and contemporary Australia, this chapter follows the advice of Patrick Wolfe that we should 'look twice' – at the other side – of anything taking place at the turn of the twentieth century, a time when 'the whole world was beginning to shift', the 'age of capitalist imperialism was dawning' and the 'horseless carriage' was coming (Wolfe, 1999: 41). Tracing the movement of certain ideas that have confined Aboriginal people in Australia, from the beginnings of the railways to the present, in what follows, the chapter articulates how such 'conceptual imprisonment' continues to enable colonial systems to profit from Aboriginal people, including imprisoned Aboriginal labour, in the interest of and for the benefit of a more mobile white population.

The chapter takes up Nicholson and Sheller's call to 'rethink how unequal relations of power inherent in both mobility and race shape a racialised mobility politics' (2016: 4) and builds on Sheller and Urry's notion that mobilities are critical to the 'marking' of racial borders and to the 'making and unmaking' of colonial and 'racial orders' (2006: 14). It responds to the question posed by Nicholson and Sheller (2016: 5): 'How can a deeper historicization of colonial and postcolonial paradigms of racial mobilities inform how we understand race and mobility today?'

At the turn of the twentieth century, the idea of Aboriginal people as savage and immoral was used to rationalise and mobilise white possession and Aboriginal containment. These racialised ideas, it is contended, persist into current times and continue to inform the rationalisation of white mobility and Aboriginal incarceration, both materially and conceptually. Taking the railways as its point of departure, the chapter investigates certain instances of the figuration and reconfiguration of these ideas that have contained Aboriginal people at both the turn of the twentieth century and in recent times. In doing so it attempts to trace the carceral logic (see also Coddington, this volume) of white possession that continues to produce such racialised ideas.

The chapter explores the bond between the mobility and immobility of settler colonisers and Aboriginal peoples, tracing how the white men who led railway developments and controlled Indigenous affairs in colonial and current times, attempt/ed to place Aboriginal people conceptually as in a state of apparent savagery and breakdown, in order to draw the Aboriginal population into useful commercial and state service. In highlighting these racialised, carceral logics, the chapter investigates the words and actions of three men notable for their engagement in the railway developments and Aboriginal affairs of their time: the First Premier of Western Australia, Sir John Forrest; his great-great nephew Andrew Forrest; and an Aboriginal railway man who speaks back to the words and actions of Sir John and Andrew Forrest, Anthony Martin Fernando.

Railway mobilities and containments

As Nicholson and Sheller (2016: 5) have observed, 'histories of transportation are crucial starting points for thinking about the relation between theories of race and mobilities in white-settler countries'. Given its colonial history and geographic spread, and the integral role it plays in the movement of people, things and ideas, the railways have always carried an association with mobility and the control of mobilities. With its particular 'material fabric and representational machinery' (Sheller and Urry, 2006: 2), rail is part of a systematised, regulated matrix in which Aboriginal peoples and settler colonisers are unevenly placed. Rail is one of many intersecting mobile systems that 'defines the contours of the modern mobilised world' (Sheller and Urry, 2006: 6) through which the prosperity and mobility of some is achieved through the control of the lands and bodies of others. While, since the turn of the twentieth century, an array of other mobility systems have developed including the motorcar, air travel, and mobile phones (Sheller and Urry, 2006), the coming of the railways is arguably a particularly useful point of departure from which to trace the Australian colonial history of how mobilities intersect with the racial and the carceral and to follow this movement into the present.

In *Tracking Modernity*, Marian Aguiar (2011) discusses how colonial thought is entangled with more celebrated ideas of modern power and technological advancement that privilege mobility. The notion of modernity has taken form through ideas of progress and expansion, dependent on the symbolic and physical movement of materials, practices, and knowledge itself. Aguiar describes how, when the colonisers developed the railway, they also introduced a venerated idea of modernity. Integral to this idea was a belief that human reason and technological advancement can overcome nature, with nature including the Aboriginal peoples who were classified alongside flora and fauna in the colonial Australian context. Colonization was and continues to be legitimized and enacted through the idea that it modernises 'uncivilised' peoples and places to a more advanced state. The railway train – a regulated and contained mechanism that compressed space and time – hastened and enabled colonial possession and expansion. It came to symbolise an idea of modernity that adhered to rationality, revered technology and privileged mobility, yet at the same time created oppressions on those who were immobilised in its creation.

The turn of the twentieth century: Fernando and Forrest

Anthony Martin Fernando was born to an Aboriginal mother in Sydney in 1864 and worked on the railways as a young man. He claimed to be the first driver of a train in a particular part of Sydney. In later years the steam engine took him to Western Australia and overseas to Europe (Paisley, 2012). Like Patrick Wolfe, Fernando also conveyed the idea of looking carefully, twice, at the other side of the technologies and ideas that forged the last century. Fernando told of how the railways, while opening up the country for white settlement, were closing down the country to his people. Despite having worked in rail himself, and relying on rail for transport and employment as a labourer and trader, Fernando warned that 'where the train goes to, that means the end for us' (Paisley, 2012: 62).

The coming of rail has been both mobilising and immobilising for Aboriginal peoples. Despite opening lines of travel, communication, and work, the movement of white people into Aboriginal land has also had an incarcerating impact which Aboriginal (Murri) poet Cec Fisher has described as 'steel rails run across my body', with Indigenous peoples and land overrun and entrapped by Western development.

> Steel rails run across my body
> These two lines of destiny
> Bring ghost people and development
> No longer aboriginal tribal grounds I see (Fisher, 1991: n.p.)

However, Sir John Forrest, the First Premier of Western Australia, had a markedly different view of the railways to that of Fernando and Fisher. Holding his premiership of Western Australian from 1890 to 1901, he became known as 'the principal political advocate for the construction of the railway' (Institute of Engineers Australia, 2001: 7) and the man who started the trans-Australian railway line (Goddard and Stannage, 1984). Greatly interested in the 'opening up' of the country for (white) settlement, mining, and development, Sir John recognised the massive potential of the railways. Upon opening the Trans-Australia railway on 17 October 1917, he claimed: 'From today, East and West are indissolubly joined together by bonds of steel, and the result must be increased prosperity and happiness for the Australian people'. But this increased prosperity and happiness was not intended to reach the First Australians, who were not entitled to enjoy this mobile life but rather subject to what Sheller and Urry (2006: 7) describe as 'movements against their will'. Trains and railways were to be put to use as 'new places and technologies that enhance the mobility of some people and places even as they also heighten the immobility of others' (Hannam et al., 2006: 3). Railways would work to create both mobilisations and moorings, with race being a deciding factor in 'what is up for grabs and what is locked in, who is able to move and who is trapped' (Hannam et al., 2006: 6).

Sir John Forrest held a vision of prosperity and happiness for (white) Australia, to be achieved through the mobilising force of 'bonds of steel'. Nevertheless, prosperous white mobility was and *is* bound to Aboriginal dispossession and containment. As Hannam et al. observe, there is no heightened mobility 'without extensive systems of immobility' (2006: 3). Mobile subjects are bound to control others who are not afforded the same relationship to movement. And as Australia's colonial history of Aboriginal 'protection', assimilation, and incarceration suggests (Baldry and Cunneen, 2014; Dodson, 2003; Fernando, 1903), such methods of control might include the removal of less mobile others, the coercion of less mobile others into serving the more mobile, or placing the less mobile in confinement.

A renowned surveyor, explorer, and pioneer, Sir John Forrest was a self-proclaimed expert on 'the Aborigines', their need for work and their detainment. When he first came to leadership, Sir John Forrest, like many colonial authorities, assumed that Aboriginal people would 'die out' and that '[g]overnments should only "do just enough to smooth their passing and to ensure that they should serve the higher civilisation before they went"' (cited in Goddard and Stannage, 1984: 56). During Sir John's rule, most Australian states were shifting from a 'protectionist' approach of 'smoothing the dying pillow' to a much more intrusive approach of close surveillance and control of Aboriginal people (Allbrook, 2014; Dodson, 2003). While in 1892 Sir John Forrest told the Natural History Society that Aboriginal people were 'a mere animal living in savagedom',[5] he later purported they 'should not be pampered but made to work ...' (Allbrook, 2014: 277).

Aileen Moreton-Robinson (2015) and Robert Williams Jr (2012) have observed that there is an 'intimate relationship between the invention of Western civilisation, the deployment of the "Savage", and the appropriation of Indigenous lands' (Moreton-Robinson, 2015: 192). By marking Indigenous people as 'savage' and white people as 'civilised' this Western idea could be used to legitimise the white takeover of Aboriginal land, to moralise the white detainment of Aboriginal bodies and to put them to work in and around the railways. As the work of Harman and Grant (2014), Green (2011), and Baldry and Cunneen (2014) explains, the excessive imprisonment of Aboriginal people is by no means new and has a violent colonial history. As 'settler' occupation of Aboriginal land expanded, 'incarceration was used as a tool to dispossess Aboriginal people' (Grant, 2015: n.p.).

As Georgine Clarsen has pointed out in the Australian settler colonial context, white occupation of Aboriginal land is

> [M]ore than just a matter of the transportation of people, things, and ideas ... onto Indigenous territories. Settler societies were constituted in, and continue to be structured by, ongoing processes of material, social, and cultural

transformation that are predicated on – expressed through and measured by – motility and mobility. Foundational to settler colonialism are both the potential and actual capacities of settlers to roam as autonomous sovereign subjects around the world and across the territories they claim as their own – and conversely *to circumscribe and control the mobilities* of Indigenous peoples, to immobilize the former sovereign owners of those territories. (Clarsen, 2015: 42)

In 1898, Sir John Forrest appointed Henry E. Prinsep as Western Australia's First Chief 'Protector of Aborigines', a role he held until 1907. Sir John regarded Prinsep as an appropriate candidate for Chief Protector given his previous work in 'efficiently set up the Mines Department', taking the view that 'he could do the same for another nascent domain of colonial government activity' (Allbrook, 2014: 281). This was a turn of the century time of mobility and confinement – when Aboriginal people were physically pushed off their land and conceptually confined by the dominant rationality of the time. In an 1899 report, Prinsep stated:

We must not forget they are savages, and we must first try to enable them to make their own work worth their food and clothing, and, if they gain this knowledge, reading and writing may then be taught; but, as they can never hope to have the same status as a white man, it is useless to teach them those things which a labourer does not require. (cited in Allbrook, 2014: 299)

Prinsep proposed that Aboriginal people be confined in missions controlled by white managers and missionaries and trained in the service of Western modernity. He praised the 'liberality' to 'their black servants' of colonisers in the South West. In August 1904, the Aborigines Act was amended to increase the powers of the Chief Protector and strengthen control over matters including employment contracts under which Aboriginal labour could be compulsorily acquired and held (Allbrook, 2014: 293).

Around this time, the colony of Western Australia came under attack, nationally and internationally, with allegations of slavery like conditions and condemnation of the colony's 'toleration of violence and forced labour' (Allbrook, 2014: 288). Sites of Aboriginal labour included but were by no means limited to the burgeoning railways under construction to facilitate the flow of white people and 'goods' across the colony. Criticism from abroad spoke out against punitive treatment including the use of neck chains, the expectation that Aboriginal workers labour without pay and forcing Aboriginal people to comply with white demands through rationing (Paisley, 2012). When in 1907 the Western Australian colony was criticised for its harsh treatment of Aboriginal peoples, Forrest responded that '[i]n no other country are aborigines looked after better ...' (Sir John Forrest, cited in *Morning Post*, 1907: 3).

As Aboriginal people were not disappearing and were not always willing to serve white people nor to face dispossession without resistance, the state was obliged to consider 'appropriate' means of confinement. According to Sir John Forrest, neck chains fitted this description. He claimed that '[c]haining aboriginals by the neck was the only effective way to prevent their escape ...' (Sir John Forrest, cited in *Morning Post*, 1907: 3). Sir John further argued that

> it is questionable whether the neck-chain is as inconvenient to the aboriginal as the wristchain, especially when he has work to do, and all native prisoners have to work on the roads near the prisons in the northern coastal towns. (Sir John Forrest, cited in *Western Mail*, 1905: 13)

Under Sir John's leadership and Prinsep's protectorship, Aboriginal prisoners endured forced labour, overcrowding, chaining and, in some cases, flogging. Indeed, Aboriginal prisoners in chains contributed to the construction of Western Australia's early railways and tramways (Martin, 1897; *West Australian Sunday Times*, 1938). Many prisoners died in custody (Green, 2011; Harman and Grant, 2014) as did Aboriginal people confined in missions.

The movement of Indigenous peoples challenges settler colonial notions of 'ordered space that should be firmly under the control of white men' who count on the control of colonised peoples and space' (Seuffert, 2011: 46). As Razack has observed: 'Indigenous people challenge the settler's claim to modernity. Is it any wonder that violence is directed at them' and 'a strenuous policing of Indigenous bodies'? (2015: 7).

The coming of the railways facilitated Fernando's travel to Western Australia in the early days of the twentieth century where he saw Aboriginal prisoners driven off their land, chained up and forced into labour. He also witnessed the murder of an Aboriginal man by white men who were never charged and was himself refused the chance to give evidence (Paisley, 2012). Horrified at the unjust imprisonment of and violence against Aboriginal Australians, Fernando petitioned Chief Protector Prinsep in a letter dated October 1903, reporting on the conditions of Aboriginal people in Western Australia. He accused missions of 'complicity with government aims to resolve finally the Aboriginal 'problem' by eliminating them altogether as a culture and people' (Paisley, 2012: 19). In his letter to Chief Protector Prinsep, Fernando stated that the mission system 'is nothing more than a moneymaking enterprise and a far worse slavery system than even the American System was' (Fernando, 1903: n.p.). He further noted that 'although the Native' is 'kicked about, whipped and thrashed' and 'made to work ... he does not receive any wages whatsoever but is almost always in a hungry and filthy state' (Fernando, 1903, cited in Paisley, 2012: 30).

While the term 'conceptual prison' is not credited to him, Fernando warned that the Aboriginal 'protection' policies in place in the early days of rail were

> never meant to cultivate the primitive mind, with a view to noble thoughts and high aspirations, or manly dignity and womanly virtue. It is only another kind of state prison, and the murder houses of the Lords and Ladies of Australasia. (Fernando, October 1903: n.p.)

Despite Fernando's reports of violence in the mission system, neither Prinsep nor Forrest acknowledged nor investigated these claims any further (Paisley, 2012: 32). As Australian authorities did not respond to his demands for justice, Fernando travelled to Europe to publicise the plight of Australia's first peoples. He gained the attention of Australian and European newspapers in the early twentieth century for his protests against 'white brutality' in Australia; protests that led to his confinement (Paisley, 2012). In perhaps his best known protest he paraded in the street outside Australia house in London in the 1920s, wearing a cloak covered in little skeletons and proclaiming, 'This is all that Australia has left of my people' (cited in Browning, 2007: n.p.).

On 2 June 1901, an article in the *West Australian Sunday Times* titled 'Black Slaves and White Brutes' condemned the brutal confinement and forced labour of Aboriginal prisoners on Western Australia's tramways. The article also notes 'the almost contemptuous attitude adopted by Sir John Forrest' in his excusing of such brutality:

> Sir John Forrest did not attempt to absolutely deny, when questioned in Melbourne, that gangs of blackfellows have been worked on the Roebourne-Cossack tramway with chains round their necks. He said he thought their hands were chained to the barrows; and that is not so. He attempted to excuse the inhumanity by saying that if they were not so chained, they would run away. Comment is unnecessary. It is evident that the place where it is necessary to chain human beings by the neck and compel them to work for paltry offences, must be nothing short of a hell upon earth. (*West Australian Sunday Times*, 1938: 11)

The Roebourne-Cossack tramway prisoners described above, and the 'Derby Aborigines in Chains' pictured in Figure 12.1 below, are among an unknown number of Aboriginal people upon whose back the Australian railway system has been built.

Incarceration – material and symbolic – continues to play a key role in enabling white occupation through the entrapping of Aboriginal people in racialised ideas and the confinement (and or elimination) of those who resist white rule. And, as the current state of Indigenous imprisonment and western expansion show, the racial dimensions of carceral mobilities remain as marked as the early days of rail.

Into the twenty-first century: The new Forrest

In 2016, the railways continue to open up land for development, whilst Indigenous Australians continue to be locked up at grossly excessive rates. What connects these apparently unrelated projects is a carceral logic that conceptually imprisons Aboriginal Australians. This conceptual imprisonment, it is argued, continues to be mobilised through the ideas informing the railway and mining development, pastoral interests, and Indigenous affairs leadership of Sir John Forrest's great-great nephew, Andrew Forrest.

Figure 12.1 Derby Aboriginals in chains, jetty
railway ballast laying, ca. 1897.

Source: State Library of Western Australia
(slwa_b2252342_1).

Andrew Forrest follows in the footsteps of his great-great uncle in many respects.
He is one of Western Australia's biggest individual landowners occupying a vast tract
of Pilbara country that the Forrest family have claimed ownership of for over 120
years. Like Sir John, Andrew Forrest benefits from the land and labour of Aboriginal
peoples who worked the property of his forbears, and from the continuing control
of the movements of traditional owners of this land. Regarded as a leading expert
on Aboriginal Affairs, in 2014 he was appointed by the Australian government to
head a major review on Indigenous jobs and training (Forrest, 2014). Aside from
his exploration and mining endeavours, he has also shown particular interest in the
work ethic of Aboriginal prisoners and crusades against global slavery.

Andrew Forrest is also a leader in railway development. He is among Australia's
richest people and Chairman of the mining corporation, Fortescue Metals Group
(FMG). In 2012 he oversaw the construction of 'the world's most efficient heavy
hall railway system' (A. Forrest, cited in FMG, 2012: n.p.) which connects his
Solomon Iron Ore mine to Port Hedland. Here again, Andrew Forrest follows
in the footsteps of his great-great uncle who 'encouraged the vast extension of
railway lines for the transport of resources (increasing the state's "prosperity"),
and for the movement of people (for colonisation to be "successful", the white
frontier needed to be advanced)' (Trotman, 2002: 99). Andrew Forrest claims to
have 'admired Lord Forrest for carrying through what he believed in' (cited in
Stewart, 2008: n.p.).

An unstated other side to Andrew Forrest's and John Forrest's visions of
dynamic growth and prosperous mobility for (white) Australia, is the containment
of the movements of Aboriginal peoples whose land's their railway lines extend
through. This containment occurs not only through the immobilising and coercive
movement of bodies but also through language and ideas. In naming the railway
he built, Andrew Forrest referred to 'the courageous and intrepid explorers of the

Pilbara', in particular his great-great aunt Hamersley, wife to Sir John Forrest. Like his great-great uncle, Andrew Forrest honours not the first peoples and traditional names of the land across which the railway runs but rather the name of his own family in his own tongue. He also continues Sir John's legacy of proclaiming himself as a friend of Aboriginal people while shackling the first peoples whose land he occupies and possesses to his own settler colonial terms and conditions. Like Sir John, Andrew Forrest is aware of the significance of both the railways – that great signifier and mobiliser of modernity and coloniality – and prisons – in the continuing possession of land.

Forrest's Fortescue Metals Group, like other leading rail and mining companies including Rio Tinto, BHP and Broadspectrum (formerly Transfield services), is promoted as offering great employment opportunities to Aboriginal people, including the promotion of programs to train and employ Aboriginal prisoners. The logic informing such programs positions white corporations as morally and rationally correct and best placed to direct the lives of Aboriginal prisoners who are positioned as in need of correction from white systems of control:

> While people are within the walls of a prison we need to do everything we can to address their offending concerns, the reason why actually they keep coming back … We need to take every single opportunity to seize that moment to correct their behaviour so we can stop them from coming back. (A. Forrest, cited in Harradine, 2014: n.p.)

This move to attract Aboriginal workers is part of broader Aboriginal engagement strategies supported by state and corporate rail and mining companies whose operations depend on agreements with Aboriginal people in the lands on which they profit. 'Mr Forrest said the jobs that would be offered to prisoners would all be in the mining industry, but could range from hospitality to heavy vehicle maintenance' (*The Western Australian*, 2014). The focus of prisoner jobs is on trainees positioned in lower level roles, based on the assumption that Aboriginal people and prisoners lack skills and training. Forrest takes pride in his motto: '[G]ive a man a fish and you have fed him for a day, teach a man to fish and you have fed him for a lifetime' (Burrell, 2013: n.p.). His statement overlooks the fact that Aboriginal people have been expert fishers for tens of thousands of years.

Another example of this broader strategy to utilise the labour of Aboriginal prisoners is the Northern Territory's 'Sentenced to a Job' programme. Under this program, Northern Territory Correctional Services have sent prisoners to labour in remote locations, what Minister John Elferink describes as 'an opportunity to change destructive patterns of behaviour through hard work' (Elferink and Styles, 2014). Unions have criticised the 'sentenced to a job' work arrangements as 'like slave labour', with large companies undercutting labour and wages to maximise profits. (Santhebennur, 2013: n.p.). Labourers receive only $60 in hand, with a portion of their earnings deducted to cover boarding and expenses and any remaining payment is put in a 'trust fund'. Given Australia's history of trust funds held for Aboriginal people and never fully paid (Kidd, 2007), questions are raised as to the destination of these moneys. The contemporary development of such prison labour programs, and the concerns they have raised around exploitative labour are reminiscent of the Aboriginal labour contracts of a century past.

Incarceration in Australia remains highly racialised, with Aboriginal people imprisoned at a rate fifteen times higher than non-Indigenous Australians (ABS, 2015; Baldry and Cunneen, 2014). Aboriginal and Torres Strait Islander prisoners accounted for just over 2 per cent of the general population, but 27 per cent of the total Australian prisoner population (ABS, 2015). Indigenous incarceration has massively increased over the last two decades (ABS, 2015; Baldry and Cunneen, 2014: 278).

Prisons too, are spaces of development and assimilation. As Jacobs has observed, the fact that Aboriginal peoples are incarcerated at far greater rates that non-Aboriginal peoples is 'far more than a socioeconomic legacy of colonialism'. It is 'a strategic instrument of assimilation' (Jacobs, 2012: i). Colonialism has not passed. As Baldry and Cunneen have indicated, there exists an 'unbroken chain from 1788 into the twenty-first century, of discriminatory institutional methods of control of Indigenous Australians, with an emphasis on various forms of detention and punishment' (2014: 291).

In 2013, Andrew Forrest joined forces with Serco, a multinational corporation simultaneously engaged in managing rail and prison operations in Australia and internationally. Former owner and manager of Australia's Great Southern Rail long distance passenger rail service, Serco sold the service in 2015, as incarceration was proving more profitable. Acacia prison in Western Australia, which boasts the largest population of Aboriginal prisoners in the state, is among the suite of immigration detention centres and prisons Serco run in Australia. Serco were given the Acacia prison management contract despite the Office of the Inspector of Custodial Services reporting that Serco have 'no experience or strategies for dealing with the large number of Aboriginal people incarcerated' (Office of the Inspector of Custodial Services, 2011: 25).

Forrest recently made a visit to Acacia prison as an act of 'faith and beyond', to inspire Indigenous inmates with his superiority and successes. The promotion of this event on Serco's website claimed:

> Having access to one of Australia's most successful men is something few people have; let alone prisoners.
>   But at Acacia Prison in Western Australia, prisoners had the opportunity to meet Andrew Forrest the man behind Fortescue Metals Group (FMG) – one of Australia's biggest mining companies as part of the Prison's 'Faith and Beyond' chaplaincy program. (Serco, 2015: n.p.)

However, what is presented as development and opportunity by Forrest and Serco may well feel like entrapment to Aboriginal people targeted for development.

As Cree Elder, Andrew Yellowback has aptly put it: 'One can die of improvement' (cited in Razack, 2015: 8). From Serco's motto of 'bringing services to life' to Acacia's 'Building a better future for everyone', rail and mining corporations now endeavor to engulf Indigenous peoples contained within prisons into their operations. What is not stated is the financial gain underlying such actions for Serco and Forrest and other corporations involved. Such efforts are part of a broader movement among leaders in mining and rail to profit from prison labour, and from the land of Aboriginal prisoners. Furthermore, such efforts are promoted as acts of generosity and faith to pull the Aboriginal prisoner out of a state of alleged offense and into useful service.

In 2014, Andrew Forrest was commissioned by Prime Minister Tony Abbott to lead a review of Indigenous jobs and training. Forrest reported on what he described as a crisis in Aboriginal disengagement with the workforce whereby Aboriginal people were not capable of managing their own affairs. His review presented getting a 'real' (e.g., mining) job as the only way to succeed. It declared: 'Jobs give individuals the opportunity to choose circumstances, to take control of their lives and to provide for their own and their children's future' (A. Forrest, 2014: 3), claiming that 'only employment will end the disparity ...' (A. Forrest, 2014: 4). Yet, Forrest's proposal is arguably more an attempt to manage and control Aboriginal people in accordance with white interests and white profit. Among the review's recommendations, for example, is a proposal that Indigenous prisoners be made to undertake previously voluntary pre-release employment programs (A. Forrest, 2014: 167). The review recommends that English classes also be made compulsory for Indigenous prisoners, while 'cultural ceremonies' are referred to in the report as 'distractions' (A. Forrest, 2014: 26). This arguably does not count as choice and control for Aboriginal people.

Forrest presents the Aboriginal communities whose land he wishes to mine as lacking training, skills and good judgement and incapable of managing their own affairs. He consequently offers to provide training and jobs in place of substantial royalties to such communities. According to Yindjibarndi traditional owner, Michael Woodley, such offers of jobs are 'just another attempt at white assimilation' and 'We don't want to be trained as labour for Fortescue's mines' (cited in Mayman, 2011: n.p.). Mining on his land would, according to Woodley, devastate his people and their land (YAC, 2011). Forrest, however, portrays the community as already devastated and positions himself as the moral, rational white man who will save 'Aboriginal people from themselves' (YAC, 2011: n.p.).

Woodley comes from the town of Roebourne, which Andrew Forrest has referred to as follows:

> If you want to join me one evening after 11 o'clock at night and walk down the streets of Roebourne and have little girls come up to you, like they have to me and offer themselves for any type of service I don't want to mention on television for the cost of a cigarette, then you know you've come to the end

of the line. Social breakdown is complete. Now I'm not going to encourage with our cash that kind of behaviour. (A. Forrest, cited in ABC, 2011: n.p.)

In a range of media reports and public statements, Andrew Forrest's pathologising of Aboriginal people drops to demoralising and offensive descriptions.

[Y]ou know you are dealing with the depths of depravity of a society that is completely broken. (A. Forrest, cited in Yakety Yak, 2011: n.p.)

In negating the morality of Aboriginal people, Andrew Forrest takes after his great-great uncle. When Chief Protector Prinsep wrote to Forrest in 1901 regarding his concern about the 'health and morals of the Aborigines', Forrest responded by underlining the word 'morals' and commenting 'they have none' (J. Forrest, cited in Allbrook, 2014: 294). As the Yindjibarndi Aboriginal Corporation (2011) put it:

It seems Mr (Andrew) Forrest needs to tell stories of 'depravity' in Aboriginal communities, so he can put himself forward as the 'humanitarian' philanthropist who will rescue Aboriginal people from themselves – all the while dispossessing traditional owners of the rights they are due from their ancestral lands. (YAC, 2011: n.p.)

Forrest further attempts to conceptually imprison Aboriginal people in a morality where crime is more sought after than education, so he can present himself as the morally and rationally upright white man who will lead Aboriginal people away from themselves and their land, clearing the way for him to take possession.

In some communities it is a badge of honour to go to jail instead of school. (A. Forrest, cited in Tillett, 2015: n.p.)

Andrew Forrest plays a pivotal role in the systems and institutions that comprise what Walter Mignolo refers to as the 'colonial matrix of power' (2011: xv). He acts to manage and confine Aboriginal people through an 'idea' that the white man is the knowing subject, 'and that Aboriginal people are different ... owing to culture, owing to frailties, pathology, anything but colonialism' (Razack, 2011: n.p.). Razack (2011) has discussed how '[t]he idea of Aboriginal difference' enables the control and confinement of Aboriginal people, such that the violence is not called violence. Aboriginal people are rendered 'as too abject, too dysfunctional, too sick, too damaged to govern themselves and consequently too damaged to own the land and govern the land' (Razack, 2011: n.p.). As Razack explains, through portraying Aboriginal peoples as 'simply pathologically unable to cope with the demands of modern life', the 'settler' secures his entitlement to the land (2015: 5). 'The settler and the settler state are both constituted as modern and as exemplary in their efforts to assist Indigenous people's entry into modernity' (Razack, 2015: 5).

Conclusion: Retracing conceptual imprisonment

Aboriginal people are incarcerated and immobilised by and through the *ideas* held about them, what Mona and Simone Tur have referred to as the 'shackles of colonial discourse' (2006: 177). This is what Lingard and Rizvi (1994: 76) describe as the 'containment of Aboriginal people by a Western colonial rationality' through which Aboriginal people have been held, 'with its coordinates "imprisoning" the colonised Other' (Lingard and Rizvi, 1994: 78). This containment takes both conceptual and material form, with, for example, the symbolic 'containment of Aboriginal people within colonial discourses of linear rationality' (Lingard and Rizvi, 1994: 79) along with the more material forms of containment in missions during the Protection era and their contemporary equivalent in the over-representation of Aboriginal people in prisons and Aboriginal deaths in custody (Lingard and Rizvi 1994). As Vine Deloria (1988: 189) has pointed out, Aboriginal peoples are 'thrown' into this prison of ideas upon which development and progress are based, and, as Welch puts it, 'still subject to colonial attitudes that ensure they remain prisoners of a Western constructed past' (2002: 34). These ideas have manifested in carceral regimes that have seen Aboriginal peoples 'put to work', ironically in creating mobile infrastructures (notably the railroad) that have systematically functioned to oppress, dispossess and immobilise them, past and present.

According to the dominant management rationality behind prisons, railways and other systems comprising the 'matrix of colonial power', 'good white men' like Andrew Forrest, are positioned as the knowing subject over incapable Indigenous others. The good white man best knows what is best for him (sic), and even when he thinks that he is 'stating what is good for "them"', he is led by what he knows best' (Mignolo, 2013: 14–15). The current state of Indigenous incarceration in Australia extends from such an idea. Excessive incarceration of colonised peoples is rationalised through the positioning of the knowing white subject as more advanced, more civilised and reasonable, against the colonised other as pathological and deficit. It is within the context of continuing colonial strategies and techniques that Indigenous people remain targets of 'carceral excess' (Baldry and Cunneen, 2014).

Railway developments have been crucial to the colonisation of Indigenous peoples and railway systems continue to operate as a site of entrapment of Aboriginal peoples. This chapter has traced how two prominent white men who have led rail developments and controlled Indigenous affairs in contemporary times and in the early days of rail have utilised the railways to produce 'racial space' and to lay the tracks for 'white control of mobility' (Nicholson and Sheller, 2016: 8). The very mobility of the Western world – as symbolised through the example of railway development and ensuing mining developments in Australia – is legitimatised by a particular construction of colonised peoples. This construction operates through racialised, carceral logics that entrap Aboriginal Australians within a conceptual prison of white rationality. The voices of Aboriginal people including activist Anthony Martin Fernando (1903), whose people the railways and missions contained, and the Yindjibarndi Aboriginal Corporation (2011), whose land

Andrew Forrest occupies and mines, inform on how this racialised logic enables mobile white bodies to profit from the conceptual and material imprisonment of Indigenous bodies. A white morality and rationality marks black bodies as immoral and irrational, thus legitimising white occupation. The current state of Indigenous imprisonment – conceptual and material – calls for a retracing of the carceral logics and a 'shifting of the geography of reasoning' (Mignolo, 2013: 14) that has enabled the excessive entrapment of Indigenous people since colonisation.

References

Aguiar M (2011) *Tracking Modernity: India's Railway and the Culture of Mobility*. London: University of Minnesota Press.

Allbrook M (2014) *Henry Prinsep's Empire: Framing a Distant Colony*. Canberra: ANU Press.

ABC (Australian Broadcasting Corporation) (2011) *Four Corners report 'Iron and Dust'*, first broadcast 18 July.

ABS (Australian Bureau of Statistics) (2015) Aboriginal and Torres Strait Islander prisoner characteristics. *4517.0 Prisoners in Australia 2015* Canberra: ABS.

Baldry E and Cunneen C (2014) Imprisoned Indigenous women and the shadow of colonial patriarchy. *Australian and New Zealand Journal of Criminology* 47(2): 276–298.

Browning D (2007) Fernando's ghost. *Hindsight* ABC Radio National, 15 July, Melbourne.

Burrell A (2013) *Twiggy: The High-Stakes Life of Andrew Forrest*. Victoria, Australia: Black Inc.

Clarsen G (2015) Special section on settler colonial mobilities. *Transfers* 5(3): 41–48.

Deloria V (1988) *Custer Died for Your Sins: An Indian Manifesto*. Norman, OK: University of Oklahoma Press.

Dodson M (2003) The end in the beginning: Re(de)finding Aboriginality. In: Grossman M (ed) *Blacklines: Contemporary Critical Writing by Indigenous Australians*, Melbourne: Melbourne University Press, 25–42.

Elferink J and Styles P (2014) *A Significant Day for Corrections in the Northern Territory*. Northern Territory Government Newsroom, 8 September 2014. Available at: www.newsroom.nt.gov.au/mediaRelease/9903

Fisher C (1991) *Unity Now*. Booval: Cec Fisher.

Fernando AM (1903) Reporting 'Cruelties towards the natives by the Government Officials in all parts of Australasia' (specifically New Norcia), Letter to H C Prinsep, President of Aborigines Board of Western Australia, Peak Hill, October 1903. State Records of Western Australia AU WA S3005- cons255 1903/0557A.

FMG (Fortescue Metals Group) (2012) *Fortescue Celebrates First Train on Fortescue Hamersley Line*. Media Release 2 December. Available at: http://fmgl.com.au/media/1685/asx-release-fortescue-celebrate-first-train-on-fortescue-hamersley-line680.pdf

Forrest A (2014) *The Forrest Review: Creating Parity*. Department of the Prime Minister and Cabinet (Australia). Available at: http://apo.org.au/node/40734

Goddard E and Stannage T (1984) John Forrest and the Aborigines. In: Reece B and Stannage T (eds) *European-Aboriginal Relations in Western Australian History. Studies in Western Australian History* 8: 52–58.

Grant E (2015) In chains: Aboriginal prisoners' bleak history. *University of Adelaide News*, 9 January. Available at: www.adelaide.edu.au/news/news75642.html

Green N (2011) Aboriginal sentencing in Western Australia in the late 19th century with reference to Rottnest Island prison. *Records of the Western Australian Museum Supplement* 79: 77–85.

Hannam K, Sheller M and Urry J (2006) Editorial: Mobilities, immobilities and moorings. *Mobilities* 1(1): 1–22.

Harman K and Grant E (2014) 'Impossible to detain … without chains'? The use of restraints on Aboriginal people in policing and prisons. *History Australia* 1(3): 157–176.

Harradine N (2014) Prisoners guaranteed work through job training scheme in WA's North West. *ABC News*, 21 November. Available at: www.abc.net.au/news/2014-11-21/guaranteed-jobs-on-release-for-prisoners-in-wa/5909132

Institute of Engineers Australia (2001) Trans-Australian Railway plaquing nomination. The Trans-Australian Railway. Nomination for award as a national engineering landmark. *WA and SA Divisions, Institute of Engineers Australia.* Available at: www.engineersaustralia.org.au/sites/default/files/Trans_Australian_Railway_Nomination.pdf

Jacobs MC (2012) Assimilation through incarceration: The geographic imposition of Canadian law over indigenous peoples. Unpublished doctoral dissertation, Queen's University, Kingston, Ontario.

Kidd R (2007) *Hard Labour, Stolen Wages: National Report on Stolen Wages.* NSW: Australians for Native Title and Reconciliation.

Lingard B and Rizvi F (1994) (Re)membering, (dis)membering: 'Aboriginality' and the art of Gordon Bennett. *Third Text* 8(26): 75–89.

Martin WHS (1897) *Derby Aboriginals in Chains, Jetty Railway Ballast Laying.* State Library of Western Australia, photograph. Available at: http://purl.slwa.wa.gov.au/slwa_b2252342_001

Mayman J (2011) This land is whose land? *Sydney Morning Herald*, April 6. Available at: www.smh.com.au/national/this-land-is-whose-land-20110405-1d30g.html#ixzz3wcw00hz1

Mignolo W (2011) *The Darker Side of Western Modernity: Global Futures, Decolonial Options.* Durham, NC: Duke University Press.

Mignolo W (2013) Epistemic disobedience, independent thought and de-colonial freedom. *Theory, Culture & Society* 26(7–8): 1–23.

Moreton-Robinson A (2015) *The White Possessive: Property, Power And Indigenous Sovereignty.* London: University of Minnesota Press.

Morning Post (1907) Treatment of Aboriginals: Sir John Forrest's opinion, 13 July, p. 3.

Nicholson JA and Sheller M (2016) Race and the politics of mobility: Introduction. *Transfers* 6(1): 4–11.

Office of the Inspector of Custodial Services (2011) *Report of an Announced Inspection of Acacia Prison.* Report No. 71, March 2011. Western Australia: Office of the Inspector of Custodial Services.

Paisley F (2012) *The Lone Protestor: A.M. Fernando in Australia and Europe.* Canberra: Aboriginal Studies Press.

Razack S (2011) Reading boot prints on the chest: Inquests into the deaths of aboriginal people in custody. *Advanced Institute for Globalisation and Culture*, 28 January, Available at: www.youtube.com/watch?v=Gv9RIIeqapM

Razack S (2015) *Dying from Improvement: Inquests and Inquiries into Indigenous Deaths in Custody.* Toronto: University of Toronto Press.

Santhebennur M (2013) Prisoners recruited on NT mine site. *Australian Mining*, 11 September, Available at: www.australianmining.com.au/news/prisonersrecruited onntminesite

Serco (2015) *Faith and Beyond.* Available at: www.serco-ap.com.au/our-services/our-innovations/faith-and-beyond/

Seuffert N (2011) Civilisation, settlers and wanderers: Law, politics and mobility in nineteenth century New Zealand and Australia. *Law Text Culture* 15: 10–44.

Sheller M and Urry J (2006) The new mobilities paradigm. *Environment and Planning A* 38(2): 207–226.

Stewart C (2008) The accidental billionaire. *The Australian*, May 24. Available at: www
.theaustralian.com.au/life/weekend-australian-magazine/the-accidental-billionaire/
story-e6frg8h6-1111116422599

Tillett A (2015) Jail a badge of honour for some Aboriginals. *The Western Australian*,
February 12. Available at: https://au.news.yahoo.com/thewest/wa/a/26269523/
jail-a-badge-of-honour-for-some-aboriginals-report/

Trotman YJ (2002) John Forrest: Western Australia under the Banyan Tree, Unpublished
doctoral dissertation, Edith Cowan University.

Tur M and Tur S (2006) Conversation: Wapar munu Mantaku Nintiringanyi: Learning about
the dreaming and land. In: Worby G and Rigney L (eds) *Sharing Spaces: Indigenous
and non-Indigenous Response Story, Country and Rights*. Perth: API Network, Australia
Research Institute, Curtin University of Technology, 160–170.

Welch C (2002) Appropriating the Didjeridu and the sweat lodge: New Age baddies and
indigenous victims? *Journal of Contemporary Religion* 17(1): 21–38.

The Western Australian (2014) Job scheme to help Aboriginal offenders. *The Western
Australian*, 25 November. Available at: https://au.news.yahoo.com/thewest/
wa/a/25590120/job-scheme-to-help-prisoners/

Western Mail (1905) The Aborigines question. 11 February, p. 13. Available at: http://nla.
gov.au/nla.news-article37382452

West Australian Sunday Times (1938) Black man talks of white brutality. 6 February, p. 11.

Williams RA Jr (2012) *Savage Anxieties: The Invention of Western Civilisation*. New York,
NJ: Palgrave Macmillan.

Wolfe P (1999) *Settler Colonialism*. London: Cassell.

YAC (Yindjibarndi Aboriginal Corporation) (2011) *Roebourne Community Demands
Apology for Forrest's Slur of Roebourne girls*. Media Release, 20 July. Available at:
http://yindjibarndi.org.au/

Yakety Yak (2011) Yakety Yak: Andrew Forrest (Apr 2011). *Centre for Social Impact,
Sydney*. Available at: www.youtube.com/watch?v=ouWyVmtdifk

13 The world of the 'rondines'

Trust, waiting, and time in a Latin American prison

Lirio Gutiérrez Rivera

Introduction

Since the 1990s, prisons in Latin America have witnessed a considerable rise in the prison population, which has led to deteriorating conditions and outbreaks of inmate violence and riots (Gutiérrez Rivera, 2013; Müller, 2012; Ungar, 2003). This not only raises questions about the traditional notions of the prison but has forced scholars to rethink the role of the prison in contemporary societies. Recent scholarship on the prison in this region has focused on understanding the impact of global policies of crime control on the prison system and the penalisation of poor male adults (Gutiérrez Rivera, 2016; Wacquant, 2003a); the exportation of penitentiary models (Müller, 2015); and prison gangs (Biondi, 2010; Gutiérrez Rivera, 2010). However, studying the Latin American prison under the lens of carceral mobilities is a project yet to be completed. Mobilities within carceral space have been garnering scholarly attention in recent years (Philo, 2014; Moran, 2015; Moran et al., 2012). Yet, most studies centre on industrial countries in the Global North. This chapter looks at mobilities in the context of Latin America, in particular a prison in Honduras. Due to lack of resources and manpower, prison administrators in Latin American prisons allow inmates to supervise and control many aspects of the prison life, such as the prison economy, food and health access, and even the movements of inmates. These self-governing regimes have become a common feature of Latin American prisons (Darke and Karam, 2016). Under these circumstances, how is inmate mobility decided? How do inmates negotiate access to move?

Based on fieldwork in the prison in Comayagua, Honduras, in 2014 and 2015, which includes visits to the prison, informal conversations with inmates and former inmates, and interviews with prison guards and former prison guards, I argue that inmate access to and restriction of movements within the prison rely on the inmates' economic capacity as well as their ability to build trust with the 'rondines'; that is, the inmates who run the self-governing regime. Analysis shows that these self-governing regimes can control the physical movement of inmates within the prison, as well as restrict the entry or transfer of certain inmates from other prisons. Ultimately, this chapter demonstrates that inmate mobility is based on power relations, which, although brutal, work to contain prison violence and riots by deciding who enters or not the prison. Conversely, it also shows that unequal access to physical movement does not necessarily mean that inmates boundaries are fixed; rather these are permeable and changing not only because of the inmates' economic capacity – and the ability to establish trust with the inmates running the self-governing regime and prison administration and staff – but also because of inmate's access (until January 2014) to certain technologies such as mobile phones.

In what follows, the next section briefly discusses the notion of mobility in the context of incarceration. The third section contextualises the contemporary prison in Latin America and the emergence of self-governing regimes. The fourth section analyses mobilities in the prison of Comayagua, Honduras, considering how the self-governing regime controls inmate movement within the prison and restricts the entry of certain inmates from other prisons. The final section provides conclusions.

Movement in the context of incarceration

The notion of mobility has generally been linked to movement, autonomy, freedom, and even transgression. As Cresswell (2006) points out, mobility did not always have these meanings. Current representations of mobility are connected to Western modernity and the expansion of capitalism as well as the rise of technologies, in particular rail transportation and, later, the automobile, and airplane. Aside from its representational meanings, mobility is also physical – that is, it is 'practiced, experienced, embodied' (Cresswell, 2006: 2). Accordingly, mobilities are not at all neutral. Moran (2015) notes that mobility is an instrument of power, as not everyone has equal access to it (see also Moran et al., 2012). Some people are excluded from it, others have total access to certain areas. In other cases, certain individuals may have restricted movement or limited access to a place or area, while others may be forced to move such as in cases of human trafficking, deportation, and detention. Here, mobilities involve global disciplinary strategies that shape mobility through technologies and spatial tactics of exclusion (Gill, 2013; Martin and Mitchelson, 2009; Mountz et al., 2012).

In the context of incarceration, prisons have generally been perceived as institutions where inmates are immobilised. As Mincke and Lemonne (2014) point out, this traditional attribute of the modern prison stems from the assumption that it not only encloses and fixes spaces, but also limits, controls, and ultimately, disciplines the movement of the inmates. The customary notion of the prison sees it as

> [A] dispositif for immobility that is painful (then rehabilitational), in the sense that punishment is the central objective for immobilising individuals in a partitioned space and cutting them off from the outside. Certain conditions have obviously been eased, but in the form of a mobility that is severely controlled by the experts based on the idea that it is possible to make individuals follow rehabilitative trajectories under strict control. Immobilisation is, thus, primal,

at the very heart of the carceral objective. Mobility is merely conditional and secondary. (Mincke and Lemonne, 2014: 533–34)

Recent scholarship has been questioning this traditional assumption of the prison. On the one hand, the idea that mobilities do exist in spaces of incarceration is gradually becoming more accepted. As Philo (2014) points out, Foucault himself explored mobilities in the prison and other 'total' institutions. On the other hand, mobilities – and not immobility – are seen as a vital part of prison life that not only shape relationships, identities, and a sense of place of the inmates and the prison staff, but also dismantle the idea of fixed boundaries – the impossible movement between the prison and the world 'outside'. For instance, Peters and Turner (2015) challenge the binary notions generally ascribed to mobility by focusing on what happens *during* movement. Not only does this show the subtleties of the politics of mobility, but also it indicates the power relationships involved in prison boundaries which are often blurred (Turner, 2014). As such, the acknowledgement of mobilities in spaces of incarceration has helped rethink the prison as a whole, as well as the relationships of the prison population, together with the boundaries within the prison and with the outside world. This new form of understanding the prison does not see it as separate and isolated place. Rather, the prison is a 'normal' place with complex relations between the prison and free society (Mincke and Lemmone, 2014: 540).

Latin America is an interesting case to explore carceral mobilities. Prisons in the region are known for their self-governing regimes, in addition to the extreme overcrowding; inmates' inhumane conditions; poor infrastructure; lack of funds to maintain the prisons and population; the constant violations of human rights; and persistent violence and massacres; among other things. Also prisons tend to mix inmates waiting to be sentenced and those already serving sentence. Furthermore, it is well known that in the past decades the prison has become one of the main sites of criminal activities, which include drug trafficking and extortion within and outside the prison (Rosen and Brienen, 2015). These facts indicate that boundaries within and outside the prison are porous; moreover, inmate mobility may also have little to do with discipline or rehabilitating an individual.

In the extreme conditions of Latin American prisons in which inmates control prison life, it is worth asking how inmates negotiate movement, as well as the circumstances in which movement is allowed and/or denied. The next section contextualises Latin American prison, paying special attention to the emergence of self-governing regimes.

The contemporary Latin American prison

The modern prison emerged in Latin America in the nineteenth century as part of the elites' nation-state and modernity projects of the recently independent countries. Inspired by liberal and modern values that granted universal rights and were believed to be necessary for the progress of society, Latin American elites imported prison models – that is, Bentham's Panopticon – from the United States and Europe with the purpose of abolishing the traditional forms of confinement and punishment established during the colonial period. However, elites were not successful in achieving the modern prison. Despite being conceived as an institution to reform and rehabilitate individuals, the prison became the site to confine poor and marginalised sectors of society – most of whom were black, indigenous, or of mixed raced. In other words, the modern prison continued to reproduce class divisions and racial constructions and imaginaries from the colonial period (Aguirre, 2005).

Similar to the late-nineteenth and early-twentieth century, the contemporary prison has little to do with the rehabilitation and reformation of individuals despite these ideas appearing in the first penal codes. Rather, the inmate population of the contemporary prison continues to come from socially and economically disadvantaged backgrounds. Throughout the twentieth century, the prison in Latin America would remain underfunded. This impeded the improvement of the prison facilities forcing prisoners to live in deplorable conditions. Because of the absence of funds, prisons have always been short of staff to supervise the inmate population. This, in turn, creates a distinct backdrop to the emergence of prisoner-regulated mobilities.

The historical difficulties of consolidating the penitentiary need to be taken into account when looking at complications occurring in the contemporary prison. On the one hand, the prison started to gain visibility in the 1990s not only because of the steady rise of the prison population, but also because of a series of tragic events such as riots, violence, and massacres in prisons in Brazil, Venezuela, and Honduras.[1] The increase of the prison population became connected to new national security policies influenced by global and regional policies of crime control, aimed at reducing crime and violence by extending prison sentences; abolishing or restricting preventive prison; and introducing militarised policing methods mainly in poor neighbourhoods (Arrarás and Bello-Pardo, 2015).

This 'penalization of poverty' (Wacquant, 2003a, 2003b) has led to considerable arrests of mostly unemployed male youth, which has had a deep impact on prison life. Overcrowding, increase of violence among inmates and between guards and inmates, inmates' lack of or limited access to health and education, and constant violations of their human rights are some of the effects of national security policies and global crime control policies on Latin American prisons (Gutiérrez Rivera, 2016). Moreover, national and global policies for crime control[2] have reshaped many aspects of prison life such as power relations, identities, and the mobility of inmates. The overall weakness of the prison institution has been its inability to deal with the huge influx of prisoners, overcrowding, and violence. In response to this persistent difficulty of supervising inmates, prison administrators have allowed or designated inmates to supervise other inmates. Self-governing regimes within the prison appear to have been always present in the prison throughout the twentieth century as a solution to the shortage of staff to regulate the prisoners. However, in the 1990s, self-governing regimes were not an alternative; rather, they were incorporated into the prison system.

Darke and Karam (2016) rightly point out that the conditions in which many inmates live in contemporary prisons are indeed inhumane and dangerous. Yet, contrary to the general assumption that these self-governing prisons are chaotic and disorganised, there *is* order and prison life *is* organised. For instance, inmates distribute food and prison resources to each other; and decide who does which chores (such as laundry, cleaning), among other things. Though sometimes crude, these self-governing regimes have helped inmates survive and, on certain occasions, have acted as buffer against prison riots and violence.

The prison in Comayagua, Honduras

In what follows I introduce the prison in Comayagua, a city located in the central part of Honduras. The prison was already in a precarious situation prior to 'Mano Dura' and conditions deteriorated after national security policies were introduced. In February 2012, one of the worst tragedies occurred. The prison caught fire leaving more than 300 inmates dead. Poor infrastructure and inmate overcrowding contributed to the high death toll. After the fire, surviving inmates helped rebuild the prison and, with the aid of the prison administration, re-established the self-governing prison order (2014, personal communication). In the next section, I explore in more detail the 'rondines', who are the inmates controlling the rest of the prisoners; the self-governing order; and its impact on inmate mobility in the Comayagua prison.

I visited the Comayagua prison in 2014 and early 2015. I interviewed prison guards, the prison administrator, and retired prison guards who had worked before the fire in February 2012. I also spoke with inmates, all of them male, aged 19–50 years. Some of the inmates had survived the fire and others arrived after it. Data collection also included interviews with former male inmates of the Comayagua prison prior to this tragedy. In general, access to any prison in Honduras is not easy, especially for females. I was able to enter the Comayagua prison thanks to a retired prison guard, who had worked there for twenty years and had established relationships with inmates, other prison guards, and a human rights organisation.

Carceral mobilities in Honduras

Prison order

Similar to the other prisons in Honduras that are overcrowded and understaffed, the prison director of the Comayagua prison allows some inmates, known as 'rondines', to organise and control prison life. As a retired prison guard and the prison director of the Comayagua prison told me, the inmate who has been in jail for the longest period is generally appointed head or 'president' of the inmates (*Presidente de los reos*) (2014, personal communication). When asked about the rationale for this choice, both answered that they believed that an extensive period of confinement had given the inmate experience and understanding about how the prison worked. Another important factor was trust in the inmate, which had been built over time.

In turn, the head or president of the inmates chooses other inmates to take positions of control. These individuals are known as the 'boss of the prison blocks' (*jefe de bartolinas*). These 'bosses' are inmates with whom the president has developed enough trust in order to hand over the supervision of the rest of prisoners in the various blocks. A 'boss' has certain status among both inmates and guards. They represent authority, autonomy, and life 'outside' the constraints of the prison. 'Bosses' enjoy more movement than the rest of the inmates. For instance, they can go back and forth between the prison where inmates are kept and the outside where guards remain. In other words, they have a 'force' (Cresswell, 2010) on movement: they can *choose* where and when to move – as well as choosing not to move – within the prison while controlling or coercing other inmates' movements.

In the self-governing prison, inmates follow a chain of command. The president and the bosses decide on various issues that organise prison life such as controlling who cooks; who helps the cook; who cleans; which economic activities are permitted – these include legal and illegal activities such as managing a food stall, selling handicrafts, and selling drugs and arms, among other things – as well as who carries them out. As a power structure, this regime subordinates and discriminates some inmates while favouring others. Poor inmates tend to experience greater restrictions for gaining access to prison resources and to movement, while those who have abilities and economic resources witness the opposite. For example, economically solvent inmates, that is, those from the middle class, are able to pay for their own cell, which is located outside of the rest of the blocks, and have privileges such as privacy, their own bed and mattress, and more space (Former inmate, 2015, personal communication). Yet middle-class inmates are a minority in the prison. Similarly, those who possess skills (such as cooking, construction, driving) or have the economic support from their families are more likely to receive benefits by prison guards and the 'bosses'. For example, a former inmate whose skills in driving trucks and construction – as well as the economic support of his family – benefitted him vis-à-vis the prison administration and the 'bosses', as he was asked on various occasions to drive trucks for deliveries. A former inmate explained that inmates from disadvantaged socio-economic backgrounds face the highest levels of discrimination and subordination because poor prisoners with few or no skills represent an economic burden for the prison administration and the 'bosses'. Thus, skills and wealth are an important aspect of inmate mobility.

Inmate mobility within the prison

In the self-governing regime, movement is not only an instrument of power; it is part of the prison economy as well as the culture of the prison. Mobilities depend on the chain of command imposed by the 'president', the relationship inmates develop with the 'bosses' and the 'president' over time, and the economic capacity of the inmate. The hierarchical structure of the self-governing regime restricts inmates' decision of location within the prison. For instance, upon arrival, inmates have to negotiate with the different bosses of the blocks their place in the blocks. When one of the bosses decides that an inmate can stay in his block, the inmate has to pay to be able to move in. The boss then assigns the inmate chores such as cleaning or laundry for the maintenance of the block. Despite securing a place within a block, inmates are at risk of being expelled from the block. The reasons for expulsion vary. However, it indicates the bosses' way of asserting power over inmate mobility through the politics of *removal*.

In some cases, the boss may prefer another inmate and will therefore get rid of an existing inmate in order to accommodate him. The new inmate may demonstrate more economic capacity and skills, or garner greater trust – an important feature in the politics of mobility as well as in the survival strategy of inmates. Trust is lost when there is tension or conflict between the boss and an inmate, which leads to the use (or abuse) of power of the boss over inmate by establishing boundaries or points of friction – the block now is off-limits and the 'boss' asserts his power over the inmate's mobility by impairing his movement in certain parts of the prison. In other cases, the boss uses crude tactics to remove an inmate such as physically and sexually abusing him in order to intimidate, humiliate, and finally remove him. Once expelled, the inmate must negotiate again with other 'bosses' of blocks to secure his accommodation. Because of the risk of being removed from a block, inmates develop various strategies in order to remain in one block, or relocate to another one if expelled. As Peters and Turner (2015) observe, in spaces of incarceration, strategies become ways for inmates to gain control over their fate and mobility. In the Comayagua prison, trust with the boss as well as with other inmates who can influence the boss or even the president becomes a common strategy of the inmates.

The development of trust between inmates, especially those who are higher up in the chain of command, takes time. However, as one inmate told me, 'time is what one has here' (Inmate, 2015, personal communication). The changes in the penal code under the hard line security policies have increased sentences and removed judges' discretionary powers. As a result, many inmates are still waiting for their sentences; that is, they are on pre-trial detention. Yet waiting comes with frustrations for the inmates. The reforms in the criminal justice system in Honduras were aimed at speeding up trials. Instead, paper work gets 'stuck', affecting the inmate who is waiting for his trial. Although the law states that each inmate should be sentenced within two years, each case can take five years or more. Thus, it is very likely that an inmate spends ten, 15, and even 20 years in prison, first, in pretrial detention and then serving his sentence.

Similar to the case of asylum seekers and migrants (see Conlon, 2011), waiting has become a normal aspect of prison life in Honduras and other Latin American countries. Waiting, as Conlon points out, is 'actively produced, embodied, experienced, politicized and resisted' (2011: 355). Auyero (2012) links waiting to power and domination. Poor people in Latin America wait long periods in their everyday encounters with state bureaucracy not simply because of the slow and sometimes inefficient bureaucracies, but because bureaucrats perceive themselves as powerful and important; thus, access to them must be regulated. Waiting, so common among the poor, is the state's way of interacting with individuals and positioning them as being unworthy of bureaucrat's time. Although waiting in prison feels eternal for some inmates who seek ways to disrupt boredom by trying to gain access to other parts of the prison,[3] this time allows inmates to develop strong ties and a level of trust with the bosses, the president, as well as with other inmates in order to not only survive in the self-governing regime, but also to influence the president's politics of mobility. Thus waiting becomes a strategy by gaining control of their fate and mobility (Peters and Turner, 2015).

The attempt to influence the president's politics of mobility is evident when inmates try to gain access to the farming area outside of the prison, which is administered jointly by the prison administration and the president. Access to this farming area, located on the outer part of the prison building, is actively sought by inmates as it provides inmates with a new set of movements or 'rhythms'; that is, 'repeated moments of movement and rest, or, alternatively, simply repeated movements with a particular measure' (Cresswell, 2010: 23). Furthermore, inmates get paid for their work on the farm. Trust is key here for influencing the politics of mobility. When there is mistrust or when there are conflicts between the inmate and the president, then the latter disables an inmate's access to the farm area as well as his earning an income.

In general, access to the farming area – and other parts outside of the area where inmates are kept – are off-limits. This includes the area where prison guards stay and where the offices are located. Prison guards, however, hardly go into the inmates' area. Even though prison guards could enter the inmate area (they have the keys to the entrances), they avoid it because of feelings of unsafety and mistrust toward inmates. Only the prison director and the sub-director will enter the inmate area and in such cases *with* the president to take the morning roll call and to select (with the bosses and president) inmates who are going to the farming area or other inmates needed for maintenance in parts of the prison. Inmates, in turn, have more restrictions gaining entry to the prison staff area not only because they are locked from the outside, but also because either they have not developed enough trust with the bosses and the president or they do not know an inmate close to the bosses or president who could recommend them.

The exception is the president of the inmates whose mobility is fluid across the various borders in the prison with no apparent *frictions* (Cresswell, 2010). He can choose where and when he wants to move; he can also decide not move at all. His mobility is not bound to economic capacity or skills but on trust with the prison director and guards. Indeed, I noticed this in my first visit to the prison:

It is the first time I come to the Comayagua prison. I sit on a bench outside the main entrance of the building while waiting for someone from the prison staff. In front of me is a monument with flowers commemorating the inmates who died in the fire in February 2012. Close to where I was sitting, there is a man counting money. He is wearing jeans, a T-shirt, and a baseball cap. He greets me and continues to count his money. One of the prison staff members approaches me and extends his hand to greet me. He then looks at the man counting the money and greets him too. This man is now talking with two other prison guards who have arrived and sat next to him. The staff member who greets me says to me, 'This is 'Muñeco', the president'. Muñeco greets me (again) and gives me a handshake. The prison staff member tells him that I am going into the prison with 'Marco', a retired prison guard. 'With "Marco"? Sure, no problem', he replies. (Field notes, August 2014)

The president not only has access to move around the whole prison grounds, he also decides who, from outside the prison, enters the inmate area. That said, the president does not interfere with family and friends who visits the inmates during the visiting hours. However, he does decide on the entry of anyone who wants to enter the prison outside of the visiting hours. In my case, I was able to enter because I was with 'Marco', a retired prison guard who had developed trust and friendship with the president, the bosses, and many inmates.

Restriction of inmate transfer from other prisons and mobile technologies

Despite the brutality that inmates at the bottom of this chain of command go through, the self-governing regime has helped to contain violence and riots in recent years. As I have explored elsewhere (Gutiérrez Rivera, 2013, 2016), prison violence and riots tend to erupt when the self-governing regime is challenged. This occurs when another self-governing regime, generally established by gang members (or maras), emerges parallel to the existing one. As a result, not only do both regimes compete brutally for space and especially the prison resources, but their coexistence changes the levels of prison violence, which now appears to be used with much more cruelty (Gutiérrez Rivera, 2016).[4]

Aware of the fact that the appearance of another self-governing regime can pose a threat to his monopoly of the prison resources, the president of the Comayagua prison managed to influence the inward mobilities of certain prisoners – that is, the transfer of some inmates from other prisons, particularly of members of the street gangs known as *maras*. In other prisons in Honduras, imprisoned mara members have challenged the existing self-governing regime by refusing to take part in it and setting up their own regime. Since 2002, when the widespread incarceration of members of the maras began, there have been various massacres involving prison guards, gang, and non-gang inmates. Two of the most brutal massacres, at El Porvenir in La Ceiba in 2003 and the San Pedro Sula prison in 2006, were directly connected to the transfer of gang member inmates from other prisons, which the president and other inmates opposed.

Prior to the fire in February 2012, the Comayagua prison received members of the maras. However, as one inmate pointed out, there were not enough to establish a self-governing regime. Nevertheless, these gang members were perceived as a potential threat. When I visited the prison in 2014 and 2015, I noticed the absence of gang members. One inmate confirmed that they were not allowed in the prison anymore after the 2012 fire. 'Only *paisas*', said one inmate, referring to those who have left the maras (Inmate, 2015, personal communication). Inmates and guards attribute the lack of prison riots and massacres to the fact that members of the maras were not allowed in this prison.

Even though the president and the bosses of the blocks control inmate mobility within the prison and even the transfer of certain inmates from other prisons for the purpose of monopolising the prison economy and resources, they have never been able to control inmate mobility involving technologies, in particular the use of mobile phones. Until January 2014, when the Honduran government prohibited the use of mobile phones in the prison in order to prevent communication between criminal gangs and drug traffickers with members outside of the prison building,[5] inmates had created a virtual space in which many said they felt 'free' because there was no regulation communicating with family, friends, spouses, girlfriends, and their children. Furthermore, the president, bosses of the blocks, and the prison administration did not interfere with this space. This changed, however, after January 2014, with the new decree.

Conclusions

This chapter has considered the politics of mobility in a Latin American prison. Focusing on the case of the prison in Comayagua, Honduras, this chapter demonstrated how inmate access and restriction to certain parts of the prison is not aimed at rehabilitating or reinserting the inmate back into society. Rather, mobility is an instrument of power that is regulated by the president of the self-governing regime. Mobilities within the prison depend on the inmate's economic capacity and skills; and trust with the president and the bosses of the block, as well as with other inmates who are well connected with these inmates at any time. An inmate's economic capacity and skills gives him benefits vis-à-vis the president and bosses: enabling him to secure accommodation and improve his access to desirable spaces around the prison site.

Trust is a key characteristic of the politics of mobility in this example. Trust with the president and the bosses allows inmates to cross prison boundaries or to remain in a block. It is also a strategy to gain control of their fate and mobility. Trust is also connected to time. The extended periods of time in the prison allow inmates to build trust with the president and bosses. Although trusted prisoners have been unable to take control of all of the workings of the prison (such as mobile phone technologies), the 'rondines' are able to maintain equilibrium in the prison by managing the intricate connections that comprise everyday life in incarceration: moving desirable prisoners around the facility, removing hostile inmates to other institutions, and thereby reducing the threat of violence in the prison.

Latin American governments have been defining the prison as an institution for the rehabilitation and social reinsertion of the individual. However, this has been impossible to achieve. Instead, self-governing regimes, which initially were an alternative to regulate inmates and prison life, are now completely incorporated into the prison system. In this context, mobilities are key to understanding how inmates negotiate and even develop resilience to survive in such harsh conditions. Because self-governing regimes appear to be a characteristic of the Latin American prison, a comparative study on carceral mobilities in the region could contribute to rethinking the role of the prison in the region and its impact on societies, and, of course, the prisoners.

Acknowledgements

This research was assisted by a grant from the Drugs, Security, and Democracy Program administered by the Social Science Research Council, in partnership with Universidad de Los Andes and Centro de Investigación Docencia y Económicas, and in cooperation with funds by the Open Society Foundations.

Notes

1 Prison violence has been on the rise since the 1990s in various Latin American prisons, which is attributed to the US 'War on Drugs' and neoliberal policies. In 1992, a prison riot in the Carandiru Penitentiary in Brazil left 111 inmates dead. In Venezuela a prison riot at the Yare penitentiary killed 25 inmates and 14 visitors. In the past decade, there have been various prison massacres in Honduras. In 2003 a prison riot at La Granja, El Porvenir, left 68 inmates dead, most of them members of one of the maras, and in 2012 a fire in the Comayagua prison left around 370 inmates dead.
2 Global policies for crime control stem from Garland's notion of the 'culture of crime control' which involves changing attitudes, beliefs, and assumptions toward punishment and criminals as a consequence of neoliberal market freedom and individual liberty as well as the dismantling of the welfare state. Today, there is a 'tough on crime' approach. Criminals, most of whom come from socially and economically disadvantaged backgrounds, are perceived as persons who made 'bad choices'. These changing attitudes toward criminals and punishment have influenced US foreign policy in Latin America. The US-led 'War on Drugs' restructured national security policies – and in turn prisons – in various countries in Latin America such as 'Mano Dura' (Iron Fist) in El Salvador and Honduras, Plan Mérida in México, and Plan Colombia.
3 It is important to note that there are no rehabilitation programs in the prison. The only organisations active in prisons are human rights and the evangelical and Catholic church.
4 There have been various prison massacres in the past fifteen years. Three of the most notorious was the massacre in 2004 at the prison El Porvenir in La Ceiba, which left 69 inmates dead, most of whom were gang members. In 2006, another massacre occurred in the prison in San Pedro Sula, killing around 160 inmates, also gang members.
5 In January 2014, the Honduran government passed the Decree 255-2013, known as the Law that Limits Mobile Services and Personal Communications in Prisons (*Ley de Limitación de Servicios de Telefonía Móvil Celular y Comunicaciones Personales (PCS) en Centros Penales a Nivel Nacional*), which states: 'Considering that prisons today have become centres for the organisation of different forms of crime, thus losing its purpose of rehabilitation and social reinsertion' (Ley de Limitación de Servicios de Telefonía Móvil Celular y Comunicaciones Personales, 2014. Decreto 255-2013, La Gaceta, Tegucigalpa), it prohibits all operators of mobile phones and personal communications from providing this service to prisons in the country.

References

Aguirre C (2005) *The Criminals of Lima and their Worlds: The Prison Experience (1850–1935)*. Durham, NC: Duke University Press.
Arrarás A and Bello-Pardo E (2015) General trends of prisons in the Americas. In: Rosen J and Brienen M (eds) *Prisons in the Americas in the Twenty-First Century. A Human Dumping Ground*. Lanham, MD: Lexington Press, 1–13.
Auyero J (2012) *Patients of the State. The Politics of Waiting in Argentina*. Durham, NC; London: Duke University Press.
Biondi K (2010) *Junto E Misturado: Uma Etnografia Do PCC*. Sao Paulo, Editora Terceiro Nome.

Cresswell T (2006) *On the Move: Mobility in the Modern Western World.* New York, NY: Routledge.

Cresswell T (2010) Towards a politics of mobility. *Environment and Planning D Society and Space* 28(1): 17–31.

Conlon D (2011) Waiting: Feminist perspectives on the spacings/timings of migrant (im) mobility. *Gender, Place and Culture* 18(3): 353–360.

Darke S and Karam ML (2016) Latin American prisons. In: Jewkes Y, Crewe B and Bennett J (eds) *Handbook on prisons*, 2nd edition, Abingdon: Routledge, 460–474.

Gill N (2013) Mobility versus liberty? The punitive uses of movement within and outside carceral environments. In: Moran D, Gill N and Conlon D (eds) *Carceral Spaces: Mobility and Agency in Imprisonment and Migrant Detention.* Farnham: Ashgate, 19–36.

Gutiérrez Rivera L (2010) Discipline and punish? Youth gangs response to zero-tolerance in Honduras. *Bulletin of Latin American Research* 29(4): 492–514.

Gutiérrez Rivera L (2013) *Territories of Violence. State, Marginal Youth, and Public Security in Honduras.* New York, NY: Palgrave.

Gutiérrez Rivera L (2016) Prison change and violence: The impact of crime control policies in Honduras. In: Howarth K and Peterson J (eds) *Crime and Violence in Latin America: Myths and Realities.* Lanham, MD: Lexington Press, 73–95.

Ley de Limitación de Servicios de Telefonía Móvil Celular y Comunicaciones Personales, 2014. Decreto 255–2013, La Gaceta, Tegucigalpa.

Martin LL and Mitchelson ML (2009) Geographies of detention and imprisonment: Interrogating spatial practices of confinement, discipline, law, and state power. *Geography Compass* 3(1): 459–477.

Mincke C and Lemonne A (2014) Prison and (im)mobility. What about Foucault? *Mobilities* 9(4): 528–549.

Moran D (2015) *Carceral Geography: Spaces and Practices of Incarceration.* Farnham: Ashgate.

Moran D, Piacentini L and Pallot J (2012) Disciplined mobility and carceral geography: Prisoner transport in Russia. *Transactions of the Institute of British Geography* 37(3): 446–460.

Mountz A, Coddington K, Catania RT and Loyd JM (2012) Conceptualizing detention: Mobility, containment, bordering, and exclusion. *Progress in Human Geography* 37(4): 522–541.

Müller MM (2012) The rise of the penal state in Latin America. *Contemporary Justice Review* 15(1): 57–76.

Müller MM (2015) Punitive entanglements: The 'war on gangs' and the making of a transnational penal apparatus in the Americas. *Geopolitics* 20(3): 696–727.

Peters K and Turner J (2015) Between crime and colony: Interrogating (im)mobilities aboard the convict ship. *Social and Cultural Geography* 16(7): 844–862.

Philo C (2014) 'One must eliminate the effects of…diffuse circulation [and] their unstable and dangerous coagulation': Foucault and beyond the stopping of mobilities. *Mobilities* 9(4): 493–511.

Rosen J and Brienen M (2015) Preface. In: Rosen J and Brienen M (eds) *Prisons in the Americas in the Twenty-First Century: A Human Dumping Ground.* Lanham, MD: Lexington Press, ix–xiv.

Turner J (2014) 'No place like home': Boundary traffic through the prison gate. In: Jones R and Johnson C (eds) *Placing the Border in Everyday Life.* Farnham: Ashgate, 227–250.

Ungar M (2003) Prisons and politics in Latin America. *Human Rights Quarterly* 25(4): 909–934.

Wacquant L (2003a) The penalization of poverty and the rise of neo-liberalism. *Capítulo Criminológico* 31(1): 7–22.

Wacquant L (2003b) Toward a dictatorship over the poor? Notes on the penalization of poverty in Brazil. *Punishment and Society* 5(2): 197–205.

Part IV

Transition

*v. a change from one form or type to another, or the
process by which this happens*

14 Enforced social mobilisation of 'deviant' women

Carceral regimes of discipline in Liverpool Female Penitentiary, 1809–1921

Kirsty Greenwood

Mobility research has been criticised over the years by feminist scholars for concentrating on men at the detriment of women's experiences (Goldthorpe, 1980). Goldthorpe, however, argues that while data *has* been produced, it has not been 'analysed as intensively' as those relating to men (1980: 277). Subsequently, 'more effort should have been devoted' to such a task (Goldthorpe, 1980: 277). This chapter goes some way in exploring and theorising previously unexploited Penitentiary data on female, semi-penal prisoners in Liverpool between 1809 and 1921 in direct relation to the physical and social mobility of their 'deviant' bodies. By 'social mobility' this chapter refers to the desired 'upwards' mobility of status for so-called 'deviant' women. However, in doing so, it attends to the distinct weakness within social mobility literature – the 'under-theorisation of mobility and power' in this context – by considering the 'coerced mobility' inherent in carceral practices (Moran et al., 2012: 446). By presenting and analysing empirical semi-penal data, this chapter calls for further inquiry into the negative aspects of that come hand-in-hand with coerced and disciplined forms of social mobilisation, drawing upon 'Foucauldian understandings of discipline and governmentality in which mobility is an instrument of power' (Moran et al., 2012: 457).

In Victorian England, women infinitely outnumbered men, and in the 1850s there were 104 females for every 100 males, or half a million 'superfluous women' (Nelson, 2007: 15). Not only were working-class women regarded as a problem, but *all* and *any* women were perceived as problematic with men questioning 'what was to be done with them?' (Neff, 1966: 1). Changes in the economic organisation of Victorian England were partly responsible for the production of 'a rigidly hierarchical and patriarchal society' with women banished from the male world of work, 'compounded by their exclusion from the dominant forms of local political and social life' (Massey, 1994: 193). Women were thus politically, physically, and socially mobilised away from public view and obscured within the family home to perform child-rearing and household duties (Cresswell, 2004). This physical mobilisation was then fortified upon the premise that any woman who deviated from her prescribed social and moral place in society, hence threatening the patriarchal social order, would be moved into a semi-penal site of reformation, such as Liverpool Female Penitentiary (hereafter LFP), UK.

The rise in the number of single women during this period accounted for the rise in deviant female behaviours such as prostitution[1] and drunkenness, which were placed on the same continuum of problematic female behaviour. In this chapter, 'deviant' women consist of those who 'did not offend against State defined criminal laws' but rather 'transgressed the strictly prescribed gender roles and deviated from social and moral norms of femininity and domesticity' (Greenwood, 2015: 18). Deviant women were physically *removed* from the family home into semi-penal spaces of reformation under the guise of social mobilisation via targeted regimes of mobility, comprising feminisation, infantilisation, and Christianisation.

The semi-penal institution was the 'third arena of social control' which 'sat between the "formal" discipline of the prison and the "informal" regulation of the domestic sphere'; depicted as the community (Barton, 2005: 17). Semi-penal institutions were not fully incarcerative due to their voluntary admission procedures, yet they adopted a punitive environment via domesticated and feminising disciplinary regimes. They were inextricably linked to wider strategies of social, biopolitical, and gendered control that targeted deviant women for semi-penal governance. Institutional governance of feminising, infantilising, and religious disciplinary regimes enabled mobility of character from 'deviant' to 'normal' woman and social mobility of disorderly woman to dutiful housewife. Whilst female prisoners were 'immobile by virtue of their imprisonment' (Moran et al., 2012: 449), as explored later, semi-penal institutionalisation in Liverpool involved an additional physical mobilisation (albeit immobilisation) of women who either contested the imposition of institutional discipline or who left fully reformed. Many were moved to other semi-penal institutions, whilst others were physically re/moved into middle-class homes for domestic service.

This research makes a unique contribution to an emerging knowledge of historical carceral mobilities. In particular, it explores the previously hidden historical female semi-penal narrative through the lens of social mobility. Although research on semi-penal institutions has been undertaken by Barton (2005) and Finnegan (1979), this study unpicks the wider biopolitical and social mobilisation efforts employed to transform deviant women into 'respectable' females between 1809 and 1921 in Liverpool. The governance of female behaviour is analysed within the family home, which ultimately warranted the physical removal of women into the LFP and subsequently, the control of female movement within the institution itself as means of normalising women's behaviour ready for a physical transition back to the family home or into a domestic position. Underpinned by a Foucauldian feminist epistemology, this chapter uncovers regimes of truth and normalisation within LFP which were engendered through the lens of social mobility. Next, is an exploration of the urban carceral landscape of Victorian Liverpool.

Carceral landscape of Victorian Liverpool

Nineteenth-century Liverpool resided in the county of Lancashire, comprising a mixture of rural villages, market towns, and cities whose industries consisted of textile manufacturing and docklands. According to Baird (2007), Liverpool's population quadrupled within the first few decades of the nineteenth century.

The growth of Liverpool as a port was mirrored by a corresponding increase in its population. Throughout the 1850s and 1860s, Liverpool was Britain's second city, holding a population of 444,000 within the borough boundary in 1861(Walton and Wilcox, 1991). This population surge, via the imposition of the Contagious Diseases Acts,[2] alerted philanthropic reformers to the increase in numbers of deviant women presenting a public health issue through the spread of venereal disease. Liverpool was 'a place where two worlds met; it was possible to pass from the prosperity of the centre of one of the world's greatest seaports to some of the most appalling slums in Europe' (Macilwee, 2006: 15). Petty criminal activity and excessive alcohol consumption were thus considered serious social problems, with the emphasis among the socially conscious on the 'improvement of the poor themselves' (Collins, 1994: 158). In addition to female drunkenness, female prostitutes soliciting on the city streets were considered both temptations to young men and as 'serving the sexual "needs" of middle-class men' (Lewis, 1986: 120). The Liverpudlian labour market consisted of unpredictable, casual work and unpredictable earnings, resulting in many working-class individuals having a 'hand-to-mouth existence' (Walton and Wilcox, 1991: 10). Matters were worsened by the limited availability of regular waged work, causing an increase in paid sexual acts outside of the family home as demonstrated in 'Saturday Night in Liverpool';

> About 50 women were charged with being idle and disorderly persons in the neighbourhood of Lime Street and London Road on Saturday night. They were arrested and taken into custody by a number of constables who were put on duty in plain clothes for this purpose. Several of the prisoners were fined 10s, and costs, being well known to the police as reputed prostitutes. (*Liverpool Mercury*, 1890: p. 5)

Whilst a configuration of deviant female behaviour, prostitution was not a criminal offence. Police, however, as is demonstrated here, often arrested women for related acts such as disorderly behaviour, which *was* defined as criminal as a concealed means of forcibly removing prostitutes from public view. Indeed, it was reported that prostitution was a moral issue tarring the city:

> It is gratifying to find, on reference to the number of brothels and prostitutes in Liverpool during the past ten years, a gradual reduction until the past year, when, as compared with the preceding year, there was an increase of 51 in the number of bad houses, and an increase of 151 in the number of prostitutes, the present numbers being- brothels 518, prostitutes 1381. These numbers however, painfully large as they are, do not convey a full idea of the extent to which prostitution prevails in Liverpool. (*Liverpool Mercury*, 1873: 5)

Police and philanthropists engaging in the physical removal of deviant women in Liverpool traces back to the early eighteenth century. Most philanthropic semi-penal institutions intended to cleanse the city of impurity and were born of knee-jerk reactions rather than methodical planning. Their objective was to forcibly remove deviant women into semi-penal spaces to undergo regimes of social mobilisation under the guise of patriarchal governance. Liverpool wanted to be 'emptied of (its) troublesome poverty' and transformed into a sin-free landscape for the enjoyment of 'consumers of urban space' (Smith, 2013: 167). It therefore operated as a porous landscape, interconnected with semi-penal institutions, police enforcement, and philanthropic ambition (Baerenholdt, 2013).

Although Simey contends that the 'selfless devotion of Victorian philanthropists to the forbidding task of building a new society in the nineteenth century was nowhere more notably demonstrated than in Liverpool' (1992:1); co-operation between philanthropic societies was lacking as they 'vied in open competition with each other for the subscription of donors' (Simey, 1992: 1; Miller, 1988). By accepting and admitting 'deviant' women, LFP anticipated halting the future supply of prostitutes. This movement of deviant female bodies from the urban cityscape into the semi-penal institution was a strategy to save the city, 'if not the nation, from moral disgrace' (Bartley, 1998: 44). Social mobility – via regimes of *removal* – was thus encapsulated in semi-penal institutional reform.

Liverpool Female Penitentiary

LFP (1809–1921), a semi-penal institution for women located at 67 Falkner Street, was one of hundreds of refuges, homes and asylums established in the UK and US in the nineteenth century. The certified objective of LFP upon its inauguration in 1809, was to provide a means for the rescue of prostitutes and 'fallen' women within Liverpool, who having followed vicious courses, were desirous of obtaining means of reformation (LFP Annual Report, 1809: 2). LFP subsequently had a dual purpose. Firstly, it operated as a conduit for the physical removal of women from the public sphere to eliminate female deviance and vice from the streets of Liverpool. The institution therefore operated *because* of mobility and the overarching desire to remove social ills from the urban landscape. Secondly, it operated to mobilise the biopolitical, social, and moral identities of deviant women, resulting in a transformation of the self. Ultimately, LFP was responsible for the physical *and* social mobilisation of deviant women in Liverpool.[3]

Moreover, LFP enabled an additional form of social mobility via its operation as a halfway house. Discharged female prisoners could receive 'shelter, moral training, and help in finding positions as domestic servants' (Rafter, 1985: 16). LFP was specifically constructed and administered to reform women identified as capable of moral improvement within an 'otherwise secular world' (Welch, 2010: 50). These carceral regimes aimed to socially mobilise deviant women, render them 'docile and useful for the material world' and prepare them 'for entry into the spiritual domain where they would meet their salvation' (Welch, 2010: 54). LFP was a distinct institute of pastoralism hidden under the guise of the operation of matron mother figures enforcing prescribed disciplinary penal regimes. Regimes of social mobility were aimed at improving physical appearance, sexual conduct, respectability, and domesticity. This emphasis upon domestic vocational training as a means of enabling social mobility is demonstrated within the Annual Report (1906);

As a rule they [female prisoners] have not been trained to any systematic work, and the aim is to teach them to be useful domestics, and after two years, if satisfactory, to provide them with an outfit and a situation. (LFP Annual Report, 1906: 5)

The moral management offered within LFP operated as one of a number of systems and forms of social obligation designed to enable the social cleansing of the city. LFP was itself a metaphor for the biopolitical reform of deviant women, concerned with how the population, at the individual level, could partake in defending society against moral threats via self-policing and self-surveillance (Baerenholdt, 2013). This reform relied on two alternate mobilities: the removal of women to the LFP and the mobilisation of new, desirable character traits.

However, a revealing article in the *Liverpool Mercury* contended that women governed *through* mobility were routinely being apprehended by police and given two options. If they conveyed visible penance for their deviance they would be sent to LFP for a period of reformation. Those who expressed little desire to reform however, were forwarded to prison for a specified period of hard labour:

The magistrates have taken action in committing to prison any fallen women found soliciting on the streets. If this law continues to be enforced it is indefensible that the public should provide homes or refuges to which they may be sent who express the wish to reform, and therefore the necessity for enlarged support to the home and similar institutions. (*Liverpool Mercury*, 1871: 6)

LFP thus represented one small piece of the governing jigsaw through which attempts at social mobility operated in Liverpool. The previous sentiment was echoed institutionally by the governing Gentlemen's Committee of LFP:

As far as we know, only three of the applications made were traceable to the recent proceedings against houses of ill-fame, a most noteworthy fact. Each of the women *professed a real desire to leave her sinful life*; they were admitted, and so far have settled quietly under the necessary rules and restrictions of the Institution. (LFP Annual Report, 1899: 4, emphasis added)

The number of women admitted by male family members increased as the nineteenth century progressed as penitentiaries were often utilised by fathers and husbands as a last attempt to 'assert some paternalistic control over their wayward females' (Barton, 2005: 69). This was representative of the 'different modalities of power', in terms of 'different ensembles of knowledge, mechanism and technique' that enabled the admission of deviant women *through* and *for* social mobility (Manderscheid et al., 2014: 483), considering 'movement *to* and *from*' the Penitentiary operated 'within a system of disciplinarity' (Moran et al., 2012: 448–449, emphasis in original), The newspaper article entitled

'Extraordinary Infatuation' demonstrates familial governance through social mobility:

> Emma Crow, a well-dressed and apparently well-educated young woman, giving her age as 24, was charged with unlawfully inter-marrying with Robert Troy on May 13, 1889, at Holy Trinity Church, Walton-on-the-Hill, Lancashire. (*Bristol Mercury*, 1890: 3)

After a period of imprisonment for bigamy Robert Troy forgave her. A number of months later, nonetheless, he found her guilty of misconduct in his home:

> Loving her very intensely, he agreed to marry her upon the condition that she would first go into a penitentiary and behave herself properly for six months. She went into a Penitentiary in Liverpool and remained there for the time specified. (*Bristol Mercury*, 1890: 3)

The 'boundary between consent and coercion' became blurred, with women being oppressed, controlled, and governed via attempts at enforced social mobility by their own family (Barton, 2005: 67). Heath explores the concept of mobility through marriage; assuming that a woman's status is 'determined by her husband's occupation' and implying that 'the women's own achievements have little to do with her husband's fate' (1981: 111). However, in the case of Emma Crow, mobility *through* marriage occurred very differently. Physical removal *into* the semi-penal institution and social mobilisation via the imposition of semi-penal discipline *within* the institution were deemed necessary for this 'deviant' woman to be reformed and mobilised into the institution of marriage.

Carceral regimes of feminisation, domestication and infantilisation

Feminisation and domestication

The disciplinary regimes evoked by the LFP Gentlemen's Committee intended to enforce social mobility and manufacture practised, docile and feminine bodies by destroying 'habits of idleness and vice' and substituting them with 'honest and profitable industry, therefore benefiting society' (Mahood, 1990: 195). Prisoners were employed in recognisably feminine tasks of sewing and weaving via reform efforts designed to mould them into respectable citizens who behaved according to the normative female role (Moran et al., 2009). Methods of feminisation and domestication represented a specific form of social mobility that demanded a transformation from a life of vice and sin to one which fitted the stereotypical image of femininity and purity. Moran et al. (2009: 706), via a Foucauldian lens, describe this feminisation as 'a disciplining power' which rendered female prisoners 'docile'. As stated in consecutive Annual Reports from 1890 to 1899:

Needlework and Washing are done in the Institution in a superior manner, and on moderate terms. Washing done at 1s, or more per dozen, according to the nature and quality of the articles sent. Needlework on moderate terms. (LFP Annual Reports, 1890–1899: various)

Additionally, as the Annual Report (1890) detailed:

The Ladies who undertook classes for instruction in reading, writing and some other elementary subjects have continued them with regularity on Friday evenings. This service is gratefully acknowledged by the Gentlemen's and Ladies' Committees; also that of the Ladies who kindly come in to read to the women while they are at work. This is much valued, as it is a help and cheer to them during their monotonous occupation of plain needlework. (LFP Annual Report, 1890: 9)

Expectations of femininity buttressed the justifications of penal discipline, with formal regulations underpinned by 'a series of informal rules which develop within a particular prison's sub-culture and which are as potent in controlling prisoners as official sanctions' (Scraton et al., 1991: 79–80).

The disciplinary regimes represented the medium whereby women were 'checked in their moral attitudes, controlled in their behaviour' and 'trained in laundry work and general household duties', ready to serve the middle-class as domestic servants; the products of successful social mobility (Bartley, 1998: 51). Indeed, there was 'an increase of work in the Laundry, and the Sewing Room has maintained its high reputation for very beautiful work' (LFP Annual Report, 1912: 11). In addition to relying upon subscriptions and donations, the expenses of the institution were achieved from the proceeds from laundry work and sewing: the predominant prisoner occupation in which inmates contributed towards their own maintenance. For middle-class women who lived in its immediate vicinity, LFP provided trained, inexpensive household help and 'reproduced the patriarchal and class order of Victorian society' by training women to become competent housekeepers (Walkowitz, 1980: 221). Deviant female identities were decisively subject to shifts within the semi-penal system of control designed to normalise them to acceptable standards of femininity and domesticity. As demonstrated in this chapter, this refers to the internalisation of mobile practices as 'governmobility' (Baerenholdt, 2013: 29). Governmobility was able to operate within the semi-penal system of control 'through bodily, technological, and institutional forms of self-government' since they were embedded within the aims, objectives and operationalisation of LFP (ibid.).

Infantilisation

Philanthropic reformers considered the female familial role as central to the Victorian moral and social order, believing that women's socialisation as mothers provided special insights in reform campaigns directed at women, whilst laying claim to goals of 'female autonomy, independence and equality' (Sangster, 2004: 232–233). The autonomy of female prisoners in LFP however, was almost non-existent, with women reduced to childlike creatures, under the moral and infanti-lising gaze of female staff:

> It is with great satisfaction that your Committee records the fact that the Evening *Classes* have begun again … These *Classes* have always been found most useful- partly in educating the *girls*, but mainly in creating a kindly interest between the young ladies conducting the *Classes* and the *girls* whom they teach. It is a well-known fact that many of the *girls* who were taught in these *Classes* in past years still retain grateful memories of their teach-ers and of the lessons learnt in the Home. (LFP Annual Report, 1904: 7, emphasis added)

Prominent within this and numerous other Annual Reports is the referral to pris-oners as *girls*. The infantilising techniques of power were very subtle; conceal-ing the genuine motivations of reformers and silencing female prisoners' already marginalised voices within official Penitentiary data. Under a normalising gaze, penitentiary staff enforced strict surveillance over female prisoners in an attempt to mobilise an internalisation of the disciplinary teachings in operation (Moran, 2015). Referring to the women as girls was a disguised mode of governance designed to expose them to full view to render them docile and malleable to reform. The Classes thus operated as an educational panopticon.

Whilst a small number of prisoners were under 16 years of age and could feasibly be classified as 'girls', it was stated in the LFP Annual Report (1918: 4) that 'the average age of those admitted this year is 19½ years, the ages rang-ing from 14 to 35'. Many women incarcerated in LFP therefore, may have been older than the 'Ladies' who were active agents in the educational Panopticon. Additionally, annual reports indicated that the women were housed in dormito-ries. The expression *dormitories*, akin to the term *girls*, echoes that of a board-ing school environment for children where female bodies were rendered docile in order for the social transformation from deviant to penitent to take place. Moreover, in 1892, outdoor summer activities dominated reforming strategies, representing a 'reduction of adult women to the status of children, depriving them of personal autonomy', and subjecting them to the 'maternal gaze' (LFP Annual Report, 1892; Barton, 2011: 95). They promoted dependency upon the institu-tion and increased the likelihood of prisoners responding with 'openness and gratitude when its personnel offered help and moral advice' (Rafter, 1983: 302). Women thus became more likely to internalise the institutional philosophy and teachings, cementing and solidifying a 'discourse of dependency' and enabling a social mobilisation to occur, moving them away from their previous devi-ance. (Haney, 2010: 40). Rather than being taught general survival skills that would promote their agency and independence, female prisoners were subject to a culture of *de*pendency which perpetuated their experiences of powerlessness (Barton, 2005; Faith, 1993).

Christianisation: An instrument of double infantilisation

Theorisations of unruly women having numerous traits in common with children dominated nineteenth century penal reform efforts. The positivist premise that a woman's moral senses were deficient; that she was revengeful, jealous, and inclined to vengeances of a refined cruelty filtered into everyday thinking (Lombroso and Ferrero, 2012). By 1890, the sanitary discourse became a location for the discussion of women's nature and for the elaboration of her social and political roles. From a religious reforming perspective, prostitution, the ultimate configuration of deviant female behaviour, was considered a moral issue 'with the burden of sin on the woman who prostitutes, not on the man who buys her services' (Faith, 1993: 24). Spiritual and moral guidance played a large role in the educative process of socially mobilising such women which supports Foucault's (1977) theory that biopower extended into the social body through the carceral archipelago, within which LFP was firmly located.

An article in the *Liverpool Mercury* (1871) stated that Christianity could cure prostitution: 'The evils and extent of prostitution in Liverpool demand the most serious and prayerful consideration of Christian ministers, of magistrates and of all interested in the religious, moral and physical condition of the community' (1871: 6). Religious ideals were hence considered a medium of ensuring social mobility. Bland contends that philanthropists drew upon Christian moral precepts to 'challenge the amorality of secular capitalist and male-dominated society' (1992: 403) whilst reconstructing the urban landscape 'as a cultural product and a cultural process, with social, cultural and political elites using their positions of power to promote their own values' (Moran, 2015: 130). For example, another article entitled 'God's rescuing hand' stated:

> Women of various ages, worn with sorrow, care and sin, with the sad-dest experiences behind them, and with only the faintest gleam of hope before them, seek shelter at its hospitable door, and there receive a kind and loving welcome. They are cheered by the hope and courage of others, taught and trained by patient and untiring perseverance, till they them-selves become strong and hopeful, and *look forward with God's blessing* to a future life of respectability and usefulness. (*Liverpool Mercury*, 1894: 4, emphasis added)

This connects to Moran's (2015) conception of the carceral cultural landscape. Although a self-governing institution, LFP utilised Christianisation to socially mobilise women's identities to promote the 'values of the State and its dominant ideologies' whilst mobilising women's 'performance via everyday practices', before release into the carceral archipelago (Moran, 2015: 130). LFP and the urban landscape of Liverpool represented '[t]he relationships of power and control out of which they have emerged, both in terms of the domain ideologies, and those elements of resistance or alternative cultures whose values may also be inscribed into them' (Moran, 2015: 130). A condition of admission was penitence: recognition of wrongdoing and eagerness to beg for forgiveness. Christianity was fundamental in receiving penance, inculcating moral principles and religious teachings, setting 'the boundaries that initially condemned women's "improper" behaviour', and establishing 'the means by which their salvation was to be achieved' (Barton, 2011: 94). Christianity was thus utilised as a vehicle for the social mobility of deviant women – encapsulated as a mode of governance.

Within the city, Massey argues that women were 'freer to escape the rigidity of patriarchal social controls' (1994: 256). However, Christianity offered a balance between urban freedom and being conservative and feminine as a woman. Because Liverpool was perceived as 'a realm of uncontrolled sexual licence' (Massey, 1994: 256), Christianity itself was mobilised as a technique to cure women of their social ills and equip them with 'personal mobility' essential for city living (Pooley et al., 2005: 13). As contended by Pooley et al., individual movement 'can itself become a performance' through which women were judged (2005: 13). Christianity inculcated self-surveillance, self-policing, and self-control – all linked to Moran's conception of urban citizens having 'constant awareness of their own visibility' and behaviour (2015: 8). Whilst religious ministers were initially reluctant to become involved in philanthropic practices of reform, Christian penal discipline was ultimately concerned with the social mobilisation and transformation of deviant female identities.

Contesting penal power

Minimal research has been undertaken to either explore specific techniques of governance *through* social mobility or semi-penal institutional methods *of* enforcing social mobilisation. This generates the task of retrieving women's authority to demonstrate their agency even when subject to physical mobilisation *into* semi-penal institutions and social mobilisation via governing semi-penal regimes. The development of regimes of mobility based upon 'domestic routine and paternalistic surveillance' was not well received by the imprisoned women (Sim, 1990: 132). The oppositional and reactive forces of disciplinary power and resistance comprised 'a fierce institutional dynamic of struggle further heightened by the physical and psychological pressures of the oppressive regime' (Carlton, 2008: 242), represented in the Annual Report (1905):

> Seven women have been restored to friends. Fourteen left of their own accord, nine were dismissed for bad conduct, eight have been sent to other Homes, and three are in Hospital, all of whom are expected to return shortly. (LFP Annual Report, 1905: 12)

Although nine prisoners were dismissed in 1905, what actually comprised 'bad conduct' to warrant dismissal is concealed within Penitentiary data. Resistance was more regular and widespread than the philanthropic gentlemen who ran the institutions would admit which accounts for its hidden nature in LFP. Women *appeared* to willingly conform to and submit to the disciplinary regimes within official data; however, akin to Barton's research on the probation hostel Vernon

Lodge, the primary sources recorded 'no incidents of serious collective insurgency and those women who were perceived to display rebellious inclinations were quickly expelled from the institution' (2005: 80). The Annual Report of LFP (1917) demonstrates this:

> Two women left of their own accord. *Five ran away*. It is a matter of deep concern to the matrons and to the members of the Committee when the women take the law into their own hands in this fashion. *It is not a prison* and everything is done to make the women conscious that their highest and best interests are suited. (LFP Annual Report, 1917: 11, emphasis added)

The non-authorised mobility of five prisoners escaping the institution contests Moran et al.'s (2012: 448) argument that 'no mobility is ever genuinely "free" or autonomous' as these prisoners were active agents in resisting the imposition of semi-penal discipline within the LFP. Resistances within or from LFP cannot thus, be conceived as isolated occurrences or attributed to the actions or pathologies of disturbed individuals (Carlton, 2008). They must be considered a part of the larger carceral archipelago and as 'a political product of institutional power relations and conditions' (Carlton, 2008: 242).

The construction of hegemony is not a one-way, top-down process; rather, it is a product of the negotiation between the dominant and the dominated – 'grounded in what Gramsci calls the "national popular"' (Gunn, 2006: 707). Hegemony is hence a site of struggle – representative in the case of a female prisoner who stole and escaped from the institution, documented in the newspaper article 'A Run Away':

> Emma Hughes, 20, was charged with absconding with clothing from the Liverpool Penitentiary. – Sergeant Lea deposed to the arrest of the prisoner at the Workhouse. Prisoner at first denied that she had been in a penitentiary in Liverpool, but afterwards admitted it – Remanded. (*Cheshire Observer*, 1874: 6)

Within a relational Foucauldian paradigm, resistance occupies a central location, playing 'the role of adversary, target, support, or handle in power relations' (Foucault, 1981: 95). Power relations are constantly in flux and subject to resistant strategies; 'within the discursive spaces that resistances create, disqualified knowledge's can be made audible' (Faith, 1993: 52).

Power relations are thus 'never seamless, but always spawning new forms of culture and subjectivity, new openings for potential resistance to emerge' (Foucault, 1983, cited in Bordo, 1993: 192). Most importantly, where there is power, there is also resistance to overt forms of control. Resistance exercised by female prisoners of LFP 'challenges prevailing discourses and delegitimizes presumptions of female inferiority' (Faith, 1993: 47). An historical semi-penal environment can 'appear orderly in terms of its regime, organisation and practices, but orderliness can be achieved through overt control and without the consent of prisoners' (Drake, 2008: 153). Whilst 'the concept of "disciplined mobility" offers a means of conceptualising mobility with limited agency" (Moran et al., 2012: 449), this research demonstrates distinct female agency via methods of prisoner resistance to the semi-penal regimes through non-authorised mobilisation. This extends Moran et al.'s argument that 'prisoners lack agency in terms of the timing, nature, route and physical circumstances' (2012: 456) to imprisonment, within a semi-penal framework which forced social mobility onto deviant women. Coerced and disciplined social mobility ultimately presented numerous opportunities for resistance within and from Liverpool Female Penitentiary.

Conclusion

Whilst a limited number of studies focus on women's historical carceral institutions, multiple areas have been neglected. Although Mahood (1990) explored the social control of women within Magdalen Homes via the use of religious reforming techniques, and Finnegan (1979) and Barton (2005) uncovered disciplinary mechanisms of power within specific semi-penal institutions, these studies have not engaged with mobilities literature to explore the methods of social, biopolitical, and physical control over deviant and criminal women in nineteenth- and early-twentieth-century England. Semi-penal institutions, rather than existing as detached spaces, possessed 'permeable boundaries and highly significant internal geographies' (Moran, 2015: 10).

Drawing upon existing mobilities and Foucauldian theorisations, this chapter has sought to weave a pattern of the distinct feminising, infantilising, and religious regimes of which deviant women were subject to in Liverpool Female Penitentiary. It has added to the limited knowledge on the 'internal spatial arrangements' of prisons in exploring the unique and under-researched *semi-penal* institution as an arena of social control in terms of 'both the institutional layout' and the 'movement of prisoners' bodies' around it in both 'space and time' (Moran, 2015: 10). It has explored how the admission of women to the Penitentiary both reflected the diverse nature of movement in and out of the city and represented a form of both physical and social mobility through enforced efforts to transform female identities from deviant to that of respectability. It has also analysed how women were subject to further 'removals' either via resistance to the imposition of specific regimes of mobility or through graduation to a domestic position at the end of their two-year sentence. The semi-penal arena subsequently personified a distinct porous space on the carceral landscape that was very distinct to that of a prison, with women free to leave if they opposed conformity to the regimes of mobility.

Whilst Baerenholdt (2013: 26) has argued for an emphasis upon 'the government *of* mobility rather than on government *through* mobility', this research has integrated both principles within the context of Liverpool and LFP to provide a unique perspective on attempts to social mobilise deviant bodies throughout the nineteenth and early twentieth century. The coexisting regimes of mobility to which women were subject to were shaped by a fusion of ideologies of dominant middle-class womanhood, which portrayed them as masculine, mad, and estranged from their roles in both the family and labour market. Although implemented and enforced by female Matrons and Sub-Matrons, the Gentlemen's Committee were in control of formulating the patriarchal mobilising penal regimes. The male judges of normality of LFP, firmly located at the top of the hierarchy of knowledge, generated the deeply gendered regimes of control and reform which served to regulate women's sexual identities and vocational bodies, thereby enabling their physical, social, and biopolitical mobilisation.

Notes

1 While the prostitute was considered the typical woman prisoner of the late nineteenth century, prostitution itself was not a criminal offence (Walkowitz, 1980). Rather, it was the social reaction to prostitution and attempts to control and regulate it which caused its stigmatisation (Walkowitz, 1980).
2 The Contagious Diseases Acts of 1864, 1866, and 1869 involved the sanitary inspection of prostitute women in an attempt to control the spread of venereal disease. All three acts constructed prostitution not as a criminal problem, but as a sexual health problem, associating it with sexually transmitted diseases and the risk of infecting healthy male members of society (Walkowitz, 1980).
3 The chief objective of LFP was to provide a temporary shelter for the rescue and reformation of women who committed moral (rather than criminal) offences. LFP operated as a means of spatial control via the 'expulsion' and 'forced removal' of deviant women from private and public life into the 'designated territory' of the Penitentiary (Moran et al., 2012: 449). As stated by Barton and Cooper (2013: 140) the semi-penal premise refers to the 'paradoxical description' of the institution in that it was 'not fully incarcerative in the sense of a prison, lacking the visible symbols of exclusionary punishment such as high walls and locked cells' but it employed a homely and domesticated, yet punitive, environment.

References

Baerenholdt J (2013) Governmobility: The powers of mobility. *Mobilities* 8(1): 20–34.
Baird C (2007) *Liverpool China Traders*. Oxford: Peter Lang.
Bartley P (1998) Preventing prostitution: The ladies association for the care and protection of young girls in Birmingham, 1887–1914. *Women's History Review* 7(1): 37–60.
Barton A (2005) *Fragile Moralities and Dangerous Sexualities: Two Centuries of Semi Penal Institutionalisation for Women*. Farnham: Ashgate.
Barton A (2011) A woman's place: Uncovering maternalistic forms of governance in the nineteenth century Reformatory. *Family and Community History* 14(2): 89–104.
Barton A and Cooper V (2013) Hostels and community justice for women: The 'semi-penal' paradox. In: Malloch M and McIvor G (eds) *Women, Punishment and Social Justice: Human Rights and Penal Practices*. Abingdon: Routledge, 136–151.
Bland L (1992) Purifying the public world: Feminist vigilantes in late Victorian England. *Women's History Review* 1(3): 397–412.
Bordo S (1993) Feminism, Foucault and the politics of the body. In: Ramazanoglu C (ed) *Up Against Foucault: Explorations of Some Tensions Between Foucault and Feminism*. Abingdon: Routledge: 179–202.
Bristol Mercury (1890) Extraordinary infatuation. *The Bristol Mercury and Daily Post*, 11 October, p. 3.
Carlton B (2008) Understanding prisoner resistance: Power, visibility and survival in high-security. In: Anthony T and Cuneen C (eds) *The Critical Criminology Companion*. Annandale: Hawkins Press, 240–252.

Cheshire Observer (1874) A run away. *Cheshire Observer*, 7 February, p. 6.

Collins N (1994) *Politics and Elections in Nineteenth-Century Liverpool*. Aldershot: Scholar Press.

Cresswell T (2004) *Place: A Short Introduction*. Oxford: Wiley-Blackwell.

Drake D (2008) Staff order in prisons. In: Bennett J, Crewe B and Wahidin A (eds) *Understanding Prison Staff*. Cullompton: Willan, 153–167.

Faith K (1993) *Unruly Women: The Politics of Confinement and Resistance*. Vancouver: Press Gang Publishers.

Finnegan F (1979) *Poverty and Prostitution: A Study of Victorian Prostitutes in York*. London: Cox and Wyman.

Foucault M (1977) *Discipline and Punish: The Birth of the Prison*. London: Penguin.

Foucault M (1981) The order of discourse. In: Young R (ed) *Untying the Text: A Post-Structuralist Reader*. Abingdon: Routledge, 48–79.

Goldthorpe J (1980) *Social Mobility and Class Structure in Modern Britain*. Oxford: Clarendon Press.

Greenwood K (2015) Applying the philosophical methodological approach of Collingwood to the Foucauldian feminist analysis of State responses to 'deviant' women in Liverpool (1809–1983). *Under Construction @ Keele* 1(2): 18–33.

Gunn S (2006) From hegemony to governmentality: Changing conceptions of power in social history. *Journal of Social History* 39(3): 705–720.

Haney L (2010) *Offending Women: Power, Punishment and the Regulation of Desire*. Berkeley, CA: University of California Press.

Heath A (1981) *Social Mobility*. Glasgow: William Collins Sons and Co.

LFP Annual Reports (1850–1918) *Annual Reports of the Liverpool Female Penitentiary*. Various titles. Liverpool: Meek, Thomas and Co. Printers.

Lewis J (1986) The working-class wife and mother and state intervention 1870–1914. In: Lewis J (ed) *Labour and Love: Women's Experience of Home and Family 1850–1940*. Oxford: Basil Blackwell, 99–122.

Liverpool Mercury (1871) The reclamation of fallen women. *Liverpool Mercury*, 14 March, p. 6.

Liverpool Mercury (1873) A year's crime in Liverpool and Manchester. *Liverpool Mercury*, 24 November, p. 5.

Liverpool Mercury (1890) Saturday night in Liverpool. *Liverpool Mercury*, 13 October, p. 5.

Liverpool Mercury (1894) The female penitentiary bazaar. *Liverpool Mercury*, 12 December, p. 4.

Lombroso C and Ferrero W (2012) The Criminal type in women. In: Muncie J, McLaughlin E and Langan M (eds) *Criminological Perspectives: Essential Readings*. London: Sage, 40–44.

Macilwee M (2006) *The Gangs of Liverpool*. Lancashire: Milo Books.

Mahood L (1990) *The Magdalenes: Prostitution in the Nineteenth Century*. Abingdon: Routledge.

Manderscheid K, Schwanen T and Tyfield D (2014) Introduction to Special Issue on Mobilities and Foucault. *Mobilities* 9(4): 479–492.

Massey D (1994) *Space, Place and Gender*. Minneapolis, MN: University of Minnesota Press.

Miller A (1988) *Poverty Deserved? Relieving the Poor in Victorian Liverpool*. Birkenhead: Liver Press.

Moran D (2015) *Carceral Geography: Spaces and Practices of Incarceration*. Farnham: Ashgate.

Moran D, Pallot J and Piacentini, L (2009). Lipstick, lace, and longing: Constructions of femininity inside a Russian prison. *Environment and Planning D: Society and Space* 27(4): 700–720.

Moran D, Piacentini L and Pallot J (2012) Disciplined mobility and carceral geography: Prisoner transport in Russia. *Transactions of the Institute of British Geographers* 37(3): 446–460.

Neff WF (1966) *Victorian Working Women: An Historical and Literary Study of Women in British Industries and Professions 1832–1850*. London: Frank Cass and Co.

Nelson C (2007) *Family Ties in Victorian England*. Westport: Greenwood Publishing Group.

Pooley C, Turnbull J and Adams M (2005) *A Mobile Century?: Changes in Everyday Mobility in Britain in the Twentieth Century*. Farnham: Ashgate.

Rafter N (1983) Chastising the unchaste: Social control functions of a women's reformatory, 1894–1931. In: Cohen S and Scull A (eds) *Social Control and the State*. Oxford: Basil Blackwell, 288–311.

Rafter N (1985) Gender, prisons and prison history. *Social Science History* 9(3): 233–247.

Sangster J (2004) Reforming women's reformatories: Elizabeth Fry, penal reform, and the state, 1950–1970. *The Canadian Historical Review* 85(2): 227–258.

Scraton P, Sim J and Skidmore P (1991) *Prisons Under Protest*. Milton Keynes: Open University Press.

Sim J (1990) *Medical Power in Prisons*. Buckingham: Open University Press.

Simey M (1992) *Charity Rediscovered: A Study of Philanthropic Effort in Nineteenth-Century Liverpool*. Liverpool: Liverpool University Press.

Smith C (2013) Spaces of punitive violence. *Criticism* 55(1): 161–168.

Walkowitz J (1980) *Prostitution and Victorian Society: Women, Class and the State*. Cambridge: Cambridge University Press.

Walton J and Wilcox A (eds) (1991) *Low Life and Moral Improvement in Mid-Victorian England*. Leicester: Leicester University Press.

Welch M (2010) Pastoral power as penal resistance: Foucault and the Groupe d'Information sur les Prisons. *Punishment & Society* 12(1): 47–63.

15 Mobilising carceral reformation
Mobility, the will to change, and the urban history of the juvenile court

Elizabeth Brown

Imprisonment, confinement, and criminal court processes are often depicted as ceasing movement and rendering stasis. Yet, movement is central to criminal court processes, and incarceration enacts what Todd Clear and his colleagues (2003) call 'coercive mobility' (also Moran et al., 2012). Instead of freedom of movement, incarceration constrains and coerces mobility, from the disruption of neighbourhood stability to the tightly controlled and highly regulated movement of incarceration. These movements include from the 'outside' to the inside of prison and jail; from points along a pathway, such as jail→courts→prisons→halfway house; between security levels such as minimum to maximum and back; between types of institutions, such as 'treatment' centres, 'reception' centres, vocational institutions and (supposedly non-carceral) civil detention spaces; and from practices of probation and parole to the negotiation of life on the 'outside' with a felony record. Instead of immobility, confinement is part of a much more fluid, mutable, and complicated series of movements across time and space (Moran et al., 2012).

Mobility, however, is not limited to movement between institutions. Rather, it is embedded *in* institutions. For example controlling mobility is central to criminal court processes whereby police use private property legal doctrines to bar individuals not convicted (or even charged) of crimes from whole city neighbourhoods (Beckett and Herbert, 2009). Criminal courts regularly issue 'stay-away' orders to people on probation and parole that extend from individuals (i.e., victims, co-defendants) to a few city blocks to an entire urban area (Rios, 2006). Civil courts even interpret nuisance laws in ways that restrict mobility when no crime is alleged and the due process protections of the criminal court are absent, such as in the case of gang injunctions (Caldwell, 2009; Stewart, 1998). Legal restraints that stop short of arresting mobility though imprisonment are important arbiters of movements, as state coercion seeks to produce, shape, and negotiate the (im) mobilities produced within these 'carceral' spaces.

Following migration scholars, detention – and by extension incarceration and institutional confinement – is not an 'end to mobility altogether' but rather 'part of a rationale to *regulate* mobility through technologies of exclusion' (Mountz et al., 2012: 5). Mobility can be desirable and progressive, as well as repressive and regulated; it can also be eased and constrained simultaneously by the same people and agencies (Blunt, 2007; Brubaker, 2010). It is the *regulation* of mobility that is central to immigration and asylum, something that continues beyond the entrance to the detention centre or the nation-state and into the mobility surveillance contained within hearing and case officer appointments (Gill, 2009). Likewise, incarceration is only one aspect of coerced mobility.

Below, I show how control of mobility and movement are central to the juvenile court. Progressive era concerns about 'slums' birthed the juvenile court, and for the past 100 years, the court played an integral role in controlling the manifestation, presentation, and mobility of entire neighbourhoods. Courts also control the mobility of individuals through practices of probation, the primary reformatory mechanism of the court, and institutionalisation. Each practice provides a further mechanism for arresting mobility in ways that 'sort' youth into typologies of those deserving and undeserving of the youthful label. Together, this history shows how the current intensification and expansion of control, surveillance, and detention are contemporary responses to larger questions about childhood mobility and lines between 'us' and 'them'.

Control and regulation of mobility in the juvenile court works with other court processes to constitute the delinquent 'other', as someone locatable, knowable, and pathological. Perhaps even more paradoxically, though, these discourses work together to partition and separate youthful bodies, casting the 'other' as simultaneously irrationally immobile (and unable or unwilling to move from spaces of criminality) and virulently mobile – infecting, terrorising, and destabilising spaces and places in ways that disrupt circulation and mobility of other networks. Migration studies of mobility demonstrate how internal and external boundaries are created and reveal the porosity of the nation-state. The lens of carceral geographies, though, provides an opportunity to glimpse how the line between internal and external is drawn absent the legal construction of citizenship.

Mobility, delinquency, and the juvenile court: A brief introduction

In 1904, Judge Ben Lindsay, one of the founders of the juvenile court movement, wrote:

> In the country or in the country town, if the boy invades the watermelon patch or the apple orchard, the neighbor can inform the father and the father can deal with the boy in the cellar or the barn in his own peculiar way. In the city the situation is entirely different. (Greene, 2003: 135)

Miriam Van Waters, a renowned 'childsaver' and court founder, declared that 'children should deal with elemental things of the world – earth, stones, trees, animals, running water, fire, open spaces – instead of pavements, signboards, subdivided lots, apartment houses, and electric percolators. Civilisation has been hardest on children' (Platt, 1977: 57). Unlike other state institutions, the juvenile court began as an institution of primarily *urban* governance in the United States.

Urban centres presented a range of dangers to children, families, and the city/nation-state. First, urban areas had 'corrupting' institutions, such as saloons, dance halls, and single, unattached men. These forces corrupted families by luring parents from their familial responsibilities and endangering the moral development of children. Critics of urban capitalism charged that waged labour took the father and mother from the hearth of the home. Juvenile courts thus developed as a mechanism to constrain the mobility of children in urban spaces and to regulate children's movement through compulsory education, curfews, home removal, and removal to rural 'training schools'.

In addition to adjudication duties, courts routinely ran detention centres, administered social service payments to families, and provided for the care of dependent children (what we would today call 'foster' children) (Rothman, 1980). Dealing with criminal behaviour was just one part of the juvenile court charge (and one that it almost never did exclusively), and as such, the range of possible governmental interventions available to the court extended beyond traditional criminal justice provision – that is, probation, imprisonment, and fining. Instead, courts across the United States employed vast power to guard against what it saw as the inevitable corrupting forces on children. These powers included probation oversight, removal from the home, imprisonment of parents for contributing to delinquency and neglect, institutional confinement, sterilisation, and subjection to the adult court. Using these powers, the court routinely exerted control over the mobility of youth in three spaces – the neighbourhood, the body, and the institution. This creates a governmentality of mobility where the court acts as an institutional node for governing movement, disciplining mobility, and enacting a sovereign territorialisation over youthful bodies.

Mobility and the neighbourhood: From 'slums' and 'ghettos' to gang injunctions

The neighbourhood is often synonymous with 'community' in juvenile justice policy, and is given extraordinary significance as one of the few 'causal' factors in delinquency. In 1912, Breckinridge and Abbot (1912) defined the 'delinquent neighbourhood', further developed by the Chicago School as 'low-income communities near the centers of commence and heavy industry [that] had the highest rates [of delinquency]' (Shaw and McKay, 1942: 3). Today, this idea continues in the 'neighbourhood effect' literature, where residential mobility is again critical to delinquency development (Sampson, 2012; Sampson and Wilson, 1995). Given the centrality of the neighbourhood to the law-crime experience, it is unsurprising that the neighbourhood was, and remains, a key site of governmental intervention.

'Delinquent neighbourhoods'

Progressive-era neighbourhoods took on criminal significance only in places where immigrant and foreign communities resided (Platt, 1977). Non-white neighbourhoods were cast as unhealthy, disease-ridden, and criminogenic spaces

(Bauman, 1987; Craddock, 2000; Shah, 2001). The Chicago School's 'delinquent neighbourhoods' were places where white middle-class norms were absent: tenements instead of single-family homes, higher presence of unmarried men, greater density, older disinvested residential structures, lack of yards, and industrial neighbours (Breckinridge and Abbott, 1912). Boarders were especially suspect for lacking familial connection and apartment houses allowed proximal contact between boarders and children (Clark, 1986). Apartments also deprived children of yards, the last vestiges of rurality in urban areas. Progressives concentrated on the neighbourhood scale reasoning that even strong families could not cope with the influence of urban degeneracy.

Juvenile courts thus sought to constrain the mobility of children in urban places. Curfew laws were commonplace, and an alone child in the city centre was enough to trigger court intervention. Juvenile curfew ordinances were first enacted in 1880, 'endorsed as a panacea [for delinquency] by President Harrison' in 1884, and numbered over 3000 in the US by 1900 (Hemmens and Bennett, 1999: 100). Seattle, like other cities, employed a special cadre of officers called the 'Purity Squad' to arrest children (and women at night) in city space and patrol illicit establishments (Putnam, 2008). Compulsory education laws mandated school attendance and children found on the street during school hours could be arrested and sent to the juvenile court. By 1918, over 70 per cent of children were in school, and by 1930, all states mandated elementary school for youth (Landes and Solmon, 1972). Juvenile courts routinely held youth for the crime of 'truancy', and specialised truancy officers even developed in some states during this time. Indeed, truancy is one of the most prolific 'status offenses' to be used throughout the history of the court to incarcerate and constrain youth, even today (Stahl, 2008). Curfews and police suppression were critical to demarcating city centres as adult space and interrupted children's ability to move through the city unencumbered by legal restrictions.

Passage of legislation restricting child labour also kept children off city streets and in schools. In Seattle, an exception to child labour laws was made for children selling newspapers. Yet, with a moral panic about newspaper vendors in the early 1910s, allowing these youth to remain on the street meant increased police surveillance, business leader oversight, and public scrutiny (Brown, 2011). That they were largely the children of immigrants meant that they were also subject to assertions about their innate criminality that largely dovetailed with eugenic racial theories.

Alongside legislative reforms, juvenile courts also sought to encourage parents to relocate from the city. One early court worker described a 'congested district which liberally sends its children into the juvenile court'. He lamented:

> Time and again the Judge has urged parents from this district to move to the numerous suburbs and gain the benefits of home life which are impossible of attainment here. But never a family among this portion of our constituency have in good spirited response folded their luggage and sought out a home beyond the din of the city. They are all with the Irishwoman, who after

a fortnight tryout in the country came bounding back into the city with kit and baggage, declaring that 'Folks is better company than stoomps'. (Seattle Juvenile Court, 1914: 32–33)

Here, several aspects of governmental 'knowledge' about the population come together. First, there are the detriments of urban living, acted upon by encouraging parents to *move* from the district (with little regard for the racial and economic restrictions that made the neighbourhood). Second, there is the knowledge about the people living in this neighbourhood in particular – the geographic references locating the neighbourhood as the place where non-white, primarily poor immigrants lived. Instead of being seen as people confronting racial and economic inequities, these residents are instead people who resist providing a 'proper' home life for children by remaining in place. For these families, their stillness is pathological, in contrast to the usual associations of safety, security, and belonging that traditionally come with 'staying put' and residential 'stability'. Together, this governmental knowledge creates a vision of 'slum' dwellers as irresponsible rational actors, drawn to depravity, and ultimately, the source of delinquency in the city.

Concerns about urban neighbourhoods in the early court developed alongside, and in many cases as a consequence of, larger concerns about urban circulation, movement, and flow. Controlling the manifestation of neighbourhoods was one way to ensure, interrupt, and produce certain types of urban areas. Through the removal of children – and (as the state hoped) entire populations – from city spaces, the state sought to bolster some types of mobility and circulation while immobilising and channelling the mobility of others. The court protected middle-class white neighbourhood mobility, while seeking to interrupt other movements, such as immigration from rural areas or settling in tenement housing. Tactics of increasing surveillance, restricting mobility, and (re)placing children that began at the inception of the court have continued to pay a critical role in subjugating, controlling, and coercing the mobility of children of colour (Feld, 1999).

Coercing community

Arresting the mobility of urban childhood continued beyond Progressives and is critical to delinquency interventions today. During World War II, curfew legislation assumed new prominence in response to the perception that children left unsupervised by war-bound fathers would succumb to the tempting city (Alvarez, 2008). Federal authorities sought to foment local efforts by establishing 'coordinating councils' in local neighbourhoods to guard against the dangers of urban America identified by the early Progressives – children in adult spaces, outside of school, and without appropriate supervision (Appier, 2005). With the advent of 'urban renewal' and other federal neighbourhood programs in the 1950s, 1960s, and 1970s, cities and localities sought comprehensive reorganisation of economically disenfranchised neighbourhoods in order to thwart delinquency. Locally based programs such as the 'Central Area Motivation Program' in 1950s Seattle and various city-based 'Model City' initiatives across the nation in the 1960s and 1970s ushered in a whole series of neighbourhood-based reforms. These initiatives targeted perceived pathologies that are remarkably consistent with the 1912 'delinquent neighbourhood'– poor, non-white, dense areas with higher rates of delinquency and greater precedence of 'slum' aesthetics. One study of delinquency even explained the persistence of youth of colour in juvenile detention by revealing that 'it was urban living, not ethnicity itself' (Department of Human Resources, 1985: 50). While the century and lexicon may have shifted, the idea that youth from delinquent neighbourhoods must have their mobility curtailed remains.

'New penology' and 'risk society' approaches to juvenile justice have further cemented the neighbourhood as a site of risk. Risk instruments, often used as a solution to racial disproportionalities, control the subjective decision of who necessitates greater risk and detention. Neighbourhood experience is an important arbiter of 'riskiness'. As I have written in another context (Brown, 2007), Seattle's risk assessment was premised on two different sets of questions: a set related to the offense and offense history and a second set related to social history, where neighbourhood experience formed the core of questions. Risky accoutrements of growing up in high-crime neighbourhoods included being friends or acquaintances with gang members, negative experiences of education, experience with drugs and alcohol, and employment history. Social and offense history scores were divided into three ratings – low, moderate, and high risk, and then plotted on a table to determine detention decision. Premised on an ideology of objectivity, offense was not even enough to trigger incarceration. Rather, only social history could trigger detention. Detainment resulted in at least two different (im)mobilisations: that which comes from being moved into confinement, and that which stems from the incredibly negative repercussions detention has for youth on future social mobility, criminal movements, and subjection to greater state controls on routines, surveillance, and state violence. Through the risk assessment, the 'distancing and marginalisation of the neighbourhood are an important element in the othering' of delinquents (Lucas, 1998: 149).

Other contemporary tactics of social control work in tandem with risk assessments and criminal courts to further control, channel, and limit the regulation and circulation of youth in the city. Neighbourhood-based tactics of social control, like broken windows and hot spots policing, are concentrated in communities of colour. Seen as a scourge on the city, enacting a virulent mobility that constrains and jeopardises the movements of others, gangs are subject to all sorts of legal techniques that transcend simple criminalisation, just as newsboys were a hundred years ago. Gang loitering ordinance and gang injunctions are just two legal mechanisms designed to interfere with youth 'sticking' to place by enforcing a mobility where youth are told to 'move along' or face prosecution in order to facilitate and foster the mobility and circulation of others (Levi, 2008; on 'stickiness', see Allen and Hollingworth, 2013). Claims that urban redevelopment and gentrification are the true causal forces for creating gang injunctions demonstrate further how competing mobilities shape the legal geographies of youth (Arnold, 2011).

Ultimately, the original juvenile court movement was 'driven by an obsessive desire to monitor, regulate, and discipline working-class and immigrant communities in the industrial city' (Greene, 2003: 139). This obsessive desire continues through 'community' oriented practices that continue to target the mobility of just a handful of urban neighbourhoods. Coercing community through the neighbourhood is, however, just one space through mobility is controlled and regulated in the juvenile court.

Mobilising reformation: Probation and institutionalisation

Practices of governance, which take population as the target, are usefully examined through the neighbourhood. However, as mobility scholars note, other governmental tactics – such as disciplinary power and territorialisation – are also critical to shaping mobility (Gill, 2009). In the juvenile court, two practices bring these other powers to the forefront – the 'softer' end of individual court intervention found in probation and the 'harder' source found in institutional confinement.

'Official friendship' and the reformable youth

Probation is the primary corrective measure employed by the court, where youth remain in the 'community' while having their movements heavily regulated, circumscribed, and surveyed. Probation in the early court was reserved for what the court considered 'normal' or 'accidental' delinquents, a practice that often continues today. It is often reserved for youth considered amenable to juvenile court intervention (often wealthier and whiter), and provides an important mediation of the line between reformation and punishment in the court.

Probation, at the turn of the twentieth century, was largely gendered, as it was mostly young men who were viewed as amenable to reform through simple surveillance in the city. Young women, by contrast, who were intimately policed in dance hall, saloons, and the city centre, especially at night, were often seen as having too much mobility. Early court reports talk of young women needing their mobility curtailed, through police supervision and institutionalisation. As one early court report put it, 'The trouble with many girls begins when their interests start to wander beyond the front gate (or more likely, perhaps, the apartment house steps). Forthwith the curious youngster who is impelled by her social impulses strikes out after them' (Seattle Juvenile Court, 1914: 25). Presence in 'dance halls', saloons, and the city centre at night were all inappropriate spaces for young women, especially those from 'apartments', as the quote suggests, and could result in juvenile court intervention and oversight (Pasko, 2010). The vast majority of young women were detained for just two crimes: incorrigibility and sexual delinquency (Knupfer, 2001). Attempts to 'protect' women's virginity often meant removal from the city altogether, and thus, probation was a much more frequent disposition for young men, revealing the gendered logics at the heart of juvenile court interventions.

Today, however, probation has substantially expanded, and instead of just a simple admonition to 'do better,' youth are instead presented with an entire litany of conditions, mandatory programming, and bodily surveillance. As a governmental technique, probation is especially critical for disciplining mobile bodies. Youth are almost always given curfews as a condition of probation, and in the San Francisco Juvenile Court, as example, this curfew is almost always set at 5.00 or 6.00 p.m. Curfews are enforced by programs that call youth at home every night and through formal questioning of parents or guardians, who are required to monitor their child for probation officers. In addition to curfews, youth are also subject to a range of conditions such as attendance at school and before- and after-school programs, counselling, community groups, probation officer check-in, electronic monitoring, and weekly urinalysis. Probation is also vast – it can encompass an entire range of conditions that might not normally be associated with criminal court processes, such as writing letters or being kind to parents.

This litany of conditions is actually a tightly controlled geography of movement that changes from day to day. Youth are required to traverse the city to make each and every court ordered appointment, and engage in court-ordered programming. Figure 15.1 illustrates just one day in the life of a probationer in San Francisco as they travel from 'early morning study academy' to high school to the range of community appointments, all before an enforced curfew of 6.00 p.m. Electronic monitoring also provides a technological innovation that allows the court to microscopically examine the movement of youth across time and space, and can tell probation officers where youth are at particular points in time. For some parts of the day, youth are rooted in one place, and then forced to move between others at more rapid rates. Finally, almost all youth are subject to some sort of stay-away order, which can cause the need to move schools, employment and travel hardships, and disconnection from intimate support networks (Beckett and Herbert, 2009).

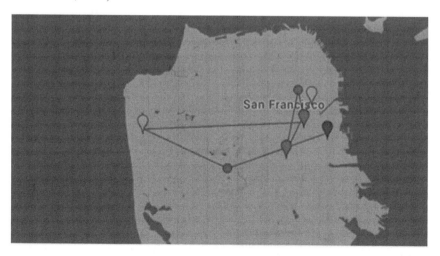

Figure 15.1 A day in the life of a probationee. (Map produced by author.)

Probation, then, is integral to shaping the intimate geographies of the city for court-adjudicated youth. Youth are exposed to unfamiliar neighbourhoods and terrain, a mobility that is often justified as liberating but is often experienced as insecure, frightening, and subject to greater danger and surveillance (Brown, 2014). It further mobilises the stigmatised identity of delinquency, by subjecting youth to 'support services' that further separate and distinguish probation youth from the non-adjudicated. Probation supervision is – and has always been – the most common court disposition (Schlossman, 1977; Torbet, 1997). However, probation is not the only way that the juvenile court acts to sort and regulate the youth population, and even today, confinement, institutionalisation, and incarceration play vital roles in regulating, controlling, and shaping the mobility of youth.

Arresting mobility through rural living

The history of reform schools is also a history of the desire to confront urban depravity and arrest the mobility of children in the city. Called industrial or training schools, reform schools are one part of the vast number of institutions that make up the juvenile court. Today and at the inception of the court, the range of residential care placements includes religious- and charitable-based group homes, secure residential placements run by private entities, county juvenile detention facilities, and congregate-based state juvenile institutions. Much of the history of the development of these types of institutions, however, is based on the idea that city living and residence itself creates delinquency and in order to intervene, the child must be removed from the city.

Visions of bucolic life, where movement is unfettered by industrial, urban, and, necessarily, degeneracy, abound in early descriptions of the ideal place of childhood. Reform schools attempted to capitalise on this notion by placing themselves within agricultural centres and attempting to create a way of life associated with an imagined pre-urban past. Rural site locations for schools resulted from the theory that something unique to urban environments caused delinquency and 'because crime was an urban phenomenon, its cure could not take place in the city' (Katz, 1968: 190). Bucolic settings were an antidote to crime, especially for the 'hardened' that necessitated removal from the city.

States around the nation lauded the idyllic attributes of institutions, often noting the amount of land, natural features, and lack of population surrounding institutions. The original Washington State Training School was placed outside the town of Chehalis, Washington, over 90 miles from Seattle, where the vast majority of imprisoned youth hailed. The first report described the school in typical bucolic terms:

> Without bolt or bar, wall or fence, a group of homelike cottages, built upon the summit of a beautiful green knoll that has been chopped out of a dense forest that surrounds on three sides, presents an inviting aspect to the prison paled child who is tremblingly led from the jail, with its bolts and bars, to what he has imagined to be a more terrible place of torment. A nearer approach and all fear is dispelled by the gladsome shout from the playground or the cheerful smile of welcome that greets him on every hand. (*First Biennial Report*, 1893: 17)

To further develop the individual along these lines, institutions were often designed to be self-sufficient. In preparing for the building of the institution, the trustees of the School wrote that they chose a 50-acre tract of land for its agricultural value, which would make the 'institution at as early a day as possible self-supporting' (*First Biennial Report*, 1893: 1). Young men at schools across the country farmed animals and plants, built structures, and ran the day-to-day operations of the school, emulating the values of rural self-sufficiency, work, and deprivation.

Early court architects largely viewed the issue of urban depravity and rural living as one of mobility. Youth were able to have the 'natural' range of movement in rural areas, where they could presumably roam the countryside unfettered by the harmful amusements of cities. Urban places, by contrast, limited the mobility of children, producing all sorts of childhood maladies like delinquency. Rural training schools thus allowed children to 'escape' the noxious influences of the city, a logic that continues today through all sorts of rural programming aimed at urban youth (Vanderbeck, 2008). After WWII, a proliferation of institutions opened that sought to further institutionalise these ideals. In the space of just a few decades, Washington State moved from just two state training schools (one was opened in 1917 for women) to a diagnostic reception centre, treatment centre, small cottage-based institution primarily designed for the youngest committed youth, and three different youth camps by 1973. Other states underwent a similar transformation. California opened two industrial schools in the late 1890s, two more for girls in 1913 and 1943, a diagnostic facility in 1942, two reception centres in 1954, and camps in 1945, 1946, 1956, 1961, 1968, and 1971. Every single one of these institutions was in a rural locale of California – Ventura (in 1913), Amador, Nevada and Calaveras counties, for instance. Further, the proliferation of camps demonstrated a strengthening commitment to rurality as ameliorating crime, but instead of the simple institution, the rural ideal was part of the physical institution itself through camping and cabin-based buildings. Many of these were located in forests, where youth practiced milling, firefighting, and other manual skills. Even urban juvenile courts opened forestry camps in surrounding areas, as California provided a subsidy for counties to open 'homes, ranches and camps' in 1945 (California Department of Corrections and Rehabilitation, 2015).

Today, coercive mobility continues to play a critical rationale in what Mountz et al. (2012) call the regulation of mobility though 'technologies of exclusion'. By removing children from the city space, the court remakes the population of urban childhood. While contemporary laws sometimes dictate that whenever possible the child should remain in the home, the decision to institutionalisation a child is nevertheless premised on the idea that the removal of children from particular spaces is necessary and good. This removal is even legislatively codified for certain crimes. While legal restrictions often mandate that children remain in the home, juvenile courts continue to remove black and brown children at far disproportionate rates (Feld, 1999). This results in a sort of reformation by geography, where removal itself is often seen as necessary to interrupt delinquent behaviour.

Reformation by geography first removes the child from the home, a first step in controlling the mobility of youth. Next, the child is often removed from the neighbourhood, as remaining in the neighbourhood is not seen as severing the ties between delinquent environments and the youth enough, as the neighbourhood is seen as 'too risky'. Finally, when youth are not successful, they are often moved to places further and further away, and if that does not work, ultimately, they are subject to the harshest coercive mobility – juvenile prison or transfer to adult court. As one probationee I interviewed put it, 'When you run, they won't help you, they just send you further'. A probation officer in a separate setting agreed, writing in a judicial report that if the youth was placed locally, it was 'a setup for failure'. Removal does not end youth mobility, but rather channels it and directs it into certain types of movements that constrain, direct, enable, and effect youth in ways not often recognised as mobilities.

Conclusion

At the advent of the juvenile court in the United States, mobility was perceived as an inherent good – the good of arresting the child from the noxious and toxic forces of urban decay, the good of removing the child from familial spaces of neglect and disrepute, and the good of regimented movement for the discipline of the body and mind of the child. Legal tactics such as youth curfews, age restrictions on business establishments, and policing of youth's presence in urban spaces are all historical examples of the juvenile court's attempt to control the mobility of youth. Mobility is as salient today as the goal of removing children from family, friends, and neighbourhoods is seen as critical to arresting the criminogenic mind and body said to develop from certain spaces, places, and experiences. The juvenile court acts as an especially important node for shaping and reshaping the mobility of youth through its decision-making practices and governmental regulation. This impact is not ephemeral and has lasting consequences over the entire life of the youth.

Analyses of mobility often neglect carceral power; yet, carceral power is integral to shaping, regulating, and reproducing logics of inequality in the United States. Likewise, carceral geographers attention to spaces of confinement often miss the wide-ranging practices of carcerality, and the ways that spaces other than confinement are implicated within governmental regimes. These types of power are integral to recreating the 'criminal class', creating categories of second-class citizenship for citizens, and for controlling mobility over the life course. Neighbourhood delinquency programs aim to control mobility at the level of population, by creating extra-ordinary surveillance, intervention, and control of just *some* urban communities. Mandated probation instantiates a coercive geographic mobility that extends the carceral complex over the entirety of the city, forcing kids into a complex and intricate network of movements seen as integral to arresting criminogenic behaviour. Institutionalisation in juvenile prisons or transfer to the adult court initiates a series of restrictions on movement reserved for the most marginalised and disenfranchised members of the populace. Though often viewed from the perspective of stasis, criminal court processes are remarkable examples of mobility in practice and are rich sites of analysis awaiting attention from scholars of mobility and carcerality.

References

Allen K and Hollingworth S (2013) 'Sticky subjects' or 'cosmopolitan creatives'? Social class, place and urban young people's aspirations for work in the knowledge economy. *Urban Studies* 50(3): 499–517.

Alvarez L (2008) *The Power of the Zoot: Youth Culture and Resistance During World War II*. Berkeley, CA: University of California Press.

Appier J (2005) 'We're blocking youth's path to crime': The Los Angeles Coordinating Councils during the Great Depression. *Journal of Urban History* 31(2): 190–218.

Arnold E (2011) Oakland gang injunctions: Gentrification or public safety? *Race, Poverty & the Environment* 18(2): 70–74.

Bauman J (1987) *Public Housing, Race, and Renewal: Urban Planning in Philadelphia, 1920–1974*. Philadelphia, PA: Temple University Press.

Beckett K and Herbert S (2009) *Banished: The New Social Control in Urban America*. New York, NY: Oxford University Press.

Blunt A (2007) Cultural geographies of migration: Mobility, transnationality and diaspora. *Progress in Human Geography* 31(5): 684–694.

Breckinridge S and Abbott E (1912) *The Delinquent Child and the Home*. New York, NY: Russell Sage Foundation Charities Publication Committee.

Brubaker R (2010) Migration, membership, and the modern nation-state: Internal and external dimensions of the politics of belonging. *Journal of Interdisciplinary History* 41(1): 61–78.

Brown E (2007) 'It's urban living, not ethnicity itself': Race, crime and the urban geography of high-risk youth. *Geography Compass* 1(2): 222–245.

Brown E (2011) The 'unchildlike child': Making and marking the child/adult divide in the juvenile court. *Children's Geographies* 9(3–4): 361–377.

Brown E (2014) Expanding carceral geographies: Challenging mass incarceration and creating a 'community orientation' towards juvenile delinquency. *Geographica Helvetica* 69(5): 377–388.

Caldwell B (2009) Criminalizing day-to-day life: A socio-legal critique of gang injunctions. *American Journal of Criminal Law* 37(3): 241–290.

California Department of Corrections and Rehabilitation (2015) *History of the DJJ*. Available at: www.cdcr.ca.gov/Juvenile_Justice/DJJ_History/Index.html

Clark CE (1986) *The American Family Home, 1800–1960*. Charlotte, NC: UNC Press Books.

Clear T, Rose D, Waring E and Scully K (2003) Coercive mobility and crime: A preliminary examination of concentrated incarceration and social disorganization. *Justice Quarterly* 20(1): 33–64.

Craddock S (2000) *City of Plagues: Disease, Poverty, and Deviance in San Francisco*. Minneapolis, MN: University of Minnesota Press.

Department of Human Resources (1985) *Preliminary Plan for At-Risk Youth*. Seattle, WA: Department of Human Resources.

Feld B (1999) *Bad Kids: Race and the Transformation of the Juvenile Court*. New York, NY: Oxford University Press.

First Biennial Report, 1891–1892: Washington State Training School (1893) Chehalis, WA: Washington State Training School.

Gill N (2009) Governmental mobility: The power effects of the movement of detained asylum seekers around Britain's detention estate. *Political Geography* 28(3): 186–196.

Greene S (2003) Vicious streets: The crisis of the industrial city and the invention of juvenile justice. *Yale Journal of Law and Human Behavior* 15(1): 135–170.

Hemmens C and Bennett K (1999) Juvenile curfews and the courts: Judicial response to a not-so-new crime control strategy. *Crime & Delinquency* 45(1): 99–121.

Katz M (1968) *The Irony of Early School Reform: Educational Innovation in Mid-Nineteenth Century Massachusetts.* New York, NY: Teachers College Press.

Knupfer A (2001) *Reform and Resistance: Gender, Delinquency, and America's First Juvenile Court.* New York, NY: Routledge.

Landes W and Solmon L (1972) Compulsory schooling legislation: An economic analysis of law and social change in the nineteenth century. *The Journal of Economic History* 32(1): 54–91.

Levi R (2008) Loitering in the city that works: On circulation, activity and police in governing urban space. In: Dubber M and Valverde M (eds) *Police and the Liberal State.* Stanford, CA: Stanford University Press, 178–202.

Lucas T (1998) Youth gangs and moral panics in Santa Cruz, California. In: Valentine G and Skelton T (eds) *Cool Places: Geographies of Youth Cultures.* New York, NY: Routledge, 145–160.

Moran D, Piacentini L and Pallot J (2012) Disciplined mobility and carceral geography: Prisoner transport in Russia. *Transactions of the Institute of British Geographers* 37(3): 446–460.

Mountz A, Coddington K, Catania RT, and Loyd J (2012) Conceptualizing detention: Mobility, containment, bordering, and exclusion. *Progress in Human Geography* 37(4): 522–541.

Pasko L (2010) Damaged daughters: The history of girls' sexuality and the juvenile justice system. *The Journal of Criminal Law and Criminology* 100(3): 1099–1130.

Platt A (1977) *The Child Savers: The Invention of Delinquency.* Chicago, IL: University of Chicago Press.

Putnam J (2008) *Class and Gender Politics in Progressive-Era Seattle.* Las Vegas, NV: University of Nevada Press.

Rios V (2006) The hyper-criminalization of Black and Latino male youth in the era of mass incarceration. *Souls* 8(2): 40–54.

Rothman D (1980) *Conscience and Convenience: The Asylum and Its Alternatives in Progressive America.* New Brunswick, NJ: Transaction Publishers.

Sampson R (2012) *Great American City: Chicago and the Enduring Neighborhood Effect.* Chicago, IL: University of Chicago Press.

Sampson R and Wilson WJ (1995) Toward a theory of race, crime, and urban inequality. In: Hagan J and Peterson R (eds) *Crime and Inequality.* Stanford, CA: Stanford University Press, 177–190.

Seattle Juvenile Court (1914) *Report for 1913.* Seattle, WA: Seattle Juvenile Court.

Schlossman S (1977) *Love & the American Delinquent: The Theory and Practice of 'Progressive' Juvenile Justice, 1825–1920.* Chicago, IL: University of Chicago Press.

Shah N (2001) *Contagious Divides: Epidemics and Race in San Francisco's Chinatown.* Berkeley, CA: University of California Press.

Shaw C and McKay H (1942) *Juvenile Delinquency and Urban Areas.* Chicago, IL: University of Chicago Press.

Stahl A (2008) Petitioned status offense cases in juvenile courts, 2004. *OJJDP Fact Sheet* 21(2) Washington, DC: US Department of Justice.

Stewart G (1998) Black codes and broken windows: The legacy of racial hegemony in anti-gang civil injunctions. *Yale Law Journal* 107(7): 2249–2279.

Torbet P (1997) Juvenile probation: The workhorse of the juvenile justice system. *Juvenile Probation Administrators Desktop Guide.* Pittsburgh, PA: National Center for Juvenile Justice.

Vanderbeck R (2008) Inner-city children, country summers: Narrating American childhood and the geographies of whiteness. *Environment and Planning A* 40(5): 1132–1150.

16 Carceral transitions experienced through Community Service placements in charity shops

Avril Maddrell

The space of the charity or thrift shop is not only a conduit for fundraising, volunteering, and alternative consumption; it is also a complex social environment that includes a variety of paid and unpaid workers. The UK charity retail sector underwent a period of professionalisation in the 1990s, including the widespread introduction of salaried shop managers and assistant managers, typically appointed on the basis of their previous retail experience. Others working in fundraising shops are generally unpaid, including volunteers who are commonly motivated to offer their time because they (a) support the mission of the charity, (b) have spare time, (c) are altruistic, (d) are seeking work experience and/ or (e) seeking social interaction. Social networks can play a significant role in drawing individuals into volunteering, e.g., through family, friends, a place of worship or care centre that has existing connections with the shop, or through social relations established through visiting the shop as a customer or donor of goods. Unpaid volunteers may also include probationers fulfilling Community Service Orders (CSOs) and prisoners on day-release licence, 'doing time' in the form of Community Service (CS)[1] where these placements have been established in particular localities between Probation and/or Prison services and the charity retail sector (Horne and Maddrell, 2002; Maddrell, 2000). This chapter focuses on Licenced Prisoners' (LPs) mobilities as experienced in and between the prison and charity shop by centring on a longitudinal study of a managed day-release placement scheme operating through a men's open prison in the South of England. The data underpinning the study is largely qualitative, including in-depth interviews with LPs, prison officers, and charity shop managers.

The use of prisoners for work is not new, with income-generating work having been seen variously as part of the punishment regime and a means of controlling prisoners (e.g., hard labour); as a means of recouping incarceration costs, enabling prisoner income-generation; and providing inmate (re)training and preparation for release. Likewise there have been numerous forms of in-prison volunteering in the UK, e.g., education or sporting facilities and programmes. Peer support volunteering within prisons includes peer-to-peer training; substance misuse support groups; language lessons; housing and other pre-release advice and support services; and Prison Listener Schemes, whereby Samaritan-trained prisoners offer befriending support to suicidal inmates (Levinson and Farrant, 2002).

These programmes, often organised in conjunction with the prison chaplaincy or other faith networks, not only support and benefit the prisoners receiving the training, support, etc., but, in common with other forms of volunteering, also improve the sense of self-worth and confidence of those volunteering (Levinson and Farrant, 2002). Additionally, there are a growing number of volunteering opportunities for non-prison beneficiaries and for selected prisoners to undertake voluntary work 'outside' of the confines of the prison.

Inmate volunteer work which benefits the non-prison population is generally limited in scope and typically involves work undertaken *within* the prison for the benefit of external organisations and their members, for example, prisoners manufacturing goods for sale by charities or offering services such as Braille transcription and repairing wheelchairs and bicycles. Some prisons also provide services within the prison to non-inmates, e.g., a scheme to assist disabled young people to exercise in the prison gym (Levinson and Farrant, 2002). All prisoner activities have to be risk-assessed; prisoners undertaking voluntary work outside of the confines of the prison have to be evaluated by a parole board and, if considered suitable, are granted a temporary licence to leave the prison for the purposes of an agreed voluntary placement. Resettlement prisons may require prisoners Released on Temporary Licence (ROTL) to undertake an *unpaid* community placement before they are eligible to receive a placement in paid work (Levinson and Farrant, 2002), as was the practice in the case study discussed here. While such obligatory schemes raise questions regarding the nature of 'volunteering' (Maddrell, 2000), prisoner-reported benefits of placements (ibid.), coupled with the very low failure rate of releases on temporary licence of 0.12 per cent (TSO 2001, cited by Levinson and Farrant, 2002) indicate that licensed-prisoner community work outside of the prison potentially offers multiple mutual benefits. However, although a small percentage of the total LPs (0.005 per cent in 2012), high-profile breaches of ROTL have resulted in tightening of regulations (Prison Blog Spot, 2015).

LPs experience of a variety of potential mobilities through Community Service placements in charity shops and their associated cost-benefits are evaluated below, after a brief outline of the nature of carceral geographies and associated mobilities.

Carceral mobilities

Within the context of recent scholarship, mobilities refer not only to movement and journeys, but also the dynamic meanings attached to and derived from these (Cresswell and Merriman, 2011). Thus, mobilities need to be recognised as embodied, imaginative, and experiential, with the capacity to co-produce the mundane and the symbolic (Maddrell, 2011, 2013). As Gale et al. note,

> Mobility/ies is more than a concept; it is a post-disciplinary paradigm that integrates leisure and tourism, transport and migration, embodied practice and performance, with the potential for new ways of seeing and thinking about these things and the links between them. (Gale et al., 2015: 7)

Attention to the nodes or fixed points, what Hannam et al. (2006) describe as the 'moorings', is also important to understanding particular mobilities. This interactive influence of co-constitutive mobilities and 'moorings' is particularly pertinent to the context of carceral mobilities where the prison and its regulatory regime constitute a supercharged mooring to which prisoners are invisibly tethered, even when outside of the confines of the prison itself.

A focus on carceral mobilities outside of the prison, such as the charity shops discussed here, highlights the ways in which organisations hosting prisoners and standing in lieu of the state can indeed be seen as fulfilling a number of shadow state functions such as surveillance and training, thereby potentially being experienced as an extension of the Panopticon functions of the prison as well as social rehabilitation (Horne and Maddrell, 2000). However, in addition to the notion of the Panopticon, Foucault's later ideas around governmentality prompt Rose (2000: 330) to argue for the recognition of 'a new archipelago of confinement' outwith the physical boundaries of detention centres, but reliant upon prisoner compliance through 'ethical self-steering mechanisms' (Rose, 2000: 321). Clearly, as Michalon notes, '[t]he spatial dimensions of confinement is not only a matter of fences' (2013: 50). This is exemplified in the case of Electronic Monitoring (EM) regimes, which are predicated on complicit self-monitoring and self-management, whereby EM simultaneously 'permits and sentences the convict to take up their social responsibility and continue to fulfil their work-related and familial obligations' (Gill, 2013: 22), as well as licensed prisoner Community Service placements that require self-regulation in order to fulfil licensing requirements and the expectations of those providing, supervising, and regulating the placements. These examples highlight both the liminal and transitional qualities of trans-carceral mobilities, i.e., the dynamic characteristics and processes experienced in such an in-between time-space, such as practising being not-a-prisoner/being in work/ social interactions (see Winnicott (1971/2005) on transitional spaces). At its core, the liminal centres on the idea of threshold, view, or entry to another life-world; but it is more than a portal. Drawing on the assemblage of spiritual mobilities that constitute pilgrimage (see Gale et al., 2015; Maddrell and della Dora, 2013; Turner and Turner, 1978), liminal qualities identified with the experience include a physical removal from one's everyday environment (though this does necessitate long-distance travel); a temporary removal of existing status; social equality; a sense of shared community and purpose; embodied social and ritual practices; renewing or transformative experiences; a sense of connection to another world or way of life; and a following return to the everyday embodying the changes experienced. Liminal-transitional characteristics may be applied in varying degrees to the experience of incarceration itself, but are equally pertinent to contexts where prisoners' mobilities cause them to segue between the confines of prison and the relative freedom of external placements or home visits.

Those charity shops that use Licensed Prisoners can be seen as dynamic permeable transcarceral spaces, which have mobilities that are emotional, imaginative, social, and physical. These mobilities are explored in this chapter through the key themes of (a) the permeable and impermeable spaces and boundaries of what constitutes prison; (b) prisoners' response to the day-to-day travel between prison and their Community Service placement, including the negotiation of moving between and operating in the different social and emotional 'worlds' of the prison and the charity shop; and (c) a critical evaluation of charity shops as a space of personal and socio-economic mobility as prisoners transition to parole, release, and rehabilitation to wider society.

Prisoner Community Service in charity shops

From the perspective of the prison service, Community Service placements potentially offer opportunities for the social rehabilitation of prisoners and a stepping stone on the path to release and resettlement, which it is hoped will reduce the likelihood of reoffending in the longer term. At the time of offending, some 75 per cent of UK convicted prisoners are unemployed; thus securing employment on release is deemed likely to reduce reoffending by up to 50 per cent (Ministry of Justice, 2012). After a full risk assessment review process, inmates are licensed to leave prison and undertake a limited range of activities under an agreed code of behaviour, typically starting with a tranche of unpaid Community Service on one or more placements, followed by paid work when close to release.

In turn, the charity retail sector gains a reliable block of hours (typically 4–5 full days per week) from usually able-bodied men aged 25–50 years, who sometimes bring skills ranging from trades such as building to financial services. Additionally, charity shop managers reported that Licensed Prisoners help increase shop sales and, perhaps contrary to expectations, improve security within shops, especially in cases of locations where shoplifting is an issue. As one shop Manager reported: 'It's wonderful to have them around security-wise … probably because of their background they're … aware of anyone dodgy. They call on the intercom for back-up or [someone else to be] watching' (Manager J, city centre shop, cited in Horne and Maddrell, 2002: 94).

Benefits for both the charity retail sector and LPs are dependent upon a positive milieu within any given shop, which both values LP input and therefore the prisoners themselves as individuals, and is able to offer skills and or curriculum vitae development to the LPs (see Maddrell, 2000; Horne and Maddrell, 2002; and Turner, 2016 on prison work). Likewise, as discussed below, LPs' own attitudes and levels of engagement are crucial to outcomes.

The findings discussed here are based on a 10-year longitudinal study, but focus primarily on the micro-, social, and symbolic mobilities identified in the most recent semi-structured in-depth interviews with 20 male Licensed Prisoners undertaking Community Service on day release from an open prison in the UK. Ninety per cent of the men were 27–50 years of age; 28 per cent had no educational qualification, while the highest qualification held was an MBA; 65 per cent were White British or Irish; 35 per cent British Black, British Asian, and Mixed Ethnicity.

Being 'outside': Physical and imaginative mobilities

Several forms of mobility were identified in Prison Service policy and LP experience. The first and perhaps most obvious mobility, which was almost universally reported by LPs as a positive experience, was simply that of being 'outside' of the confines of the prison, which for some allowed at least a temporary imaginative time-space to 'forget' being in prison (and arguably being a prisoner) (see also Gacek, this volume, on forgetting):

> [Being on the placement was] a lot better … you are going to enjoy being out more than being in … you are with different people aren't you, in here you are with the same people every day, it is just the same old rubbish, but out there you are with normal people just interacting, yes, just living a normal life again it seems, you start forget that you are in prison really. (LP M, Aged 25–34, Mixed Ethnicity)

That said, another respondent highlighted how the relative pleasures and freedoms in the space and community of the placement were relational to the necessity of returning to prison which impinged upon and *tethered* his experience: 'it's a closer reminder that you are still in prison, even though you are out of the prison' (LP P, Aged 25–34, Mixed Ethnicity).

The actual travel to and from placements also provided insights to different forms of physical-social mobility. One LP described the pleasure of 'driving through the countryside' to his placement; another found little scope for personal development through his placement as a prison van driver taking others to and from their placements, but experienced the micro-space of the van itself a space of relative autonomy, enjoying: 'a bit of freedom and putting the radio on' (LP X, Aged 45–54, White British). In contrast, for another, a former white-collar professional convicted of fraud, the journey to the placement within the confined space and proximity of the van forced him into close contact with other prisoners. This proximity underscored his own status as prisoner and consequent lack of autonomy: '[I]t would be nice to have been able to make our own way there and back to be frank, a load of other prisoners in a minibus is not the best start to every day' (LP E, Aged 35–44, White British).

Social mobilities and the new normal: Transitioning towards release

Social mobility, in the sense of moving within different social groups and adapting socially, was a key characteristic of reported LP experience. As the following respondent articulates, social interactions through Community Service placements facilitate a gradual and managed reintroduction to non-prison social spaces and relations, compared to the 'shock' of immediate release from prison:

> Gives you that bit of freedom doesn't it, and gives you the chance to actually mingle back in, so it was good. Because I think a lot of people, they've been in custody and they get thrown out, that's a big shock, and with this it's

more of a gradual process, which is all right … If you get just released it's a bit different, but with doing the community work and that it introduces you doesn't it, introduces you back into the community at a nice steady pace. (LP R, Aged 35–44, Black British)

Another aspect of this transitional liminal space constituted by Community Service placements is that of relative 'normality', *if* LPs are socially accepted within the space and community of the shop. While this experience was not universal, where social acceptance of LPs occurred, it impacted both on LP well-being and their attitude to the Community Service work:

So it was a good thing. I thought it could have been a waste of time but it wasn't actually, it was quite a shock … It was all a good experience … I enjoyed it, it was just getting on with everyone, everyone treated me normal, it wasn't like it made no difference where I was from, I thought that would have been an issue but it wasn't. No, they made me feel welcome, it was fine really. (LP M, Aged 25–34, Mixed Ethnicity)

Another young LP, convicted for robbery, stressed his pleasure in interacting with older female volunteers who reminded him of his grandmother, and of his sense of his own role as providing youthful entertainment for them:

Right, well I like to help people, I help the elderly, there's a lot of elderly ladies that take time out of their schedule to go and do a bit, so you know I have good conversations with them and you know, they remind me of my grandmother. So I get quite a lot of enjoyment out of that.

So is that the social side [of the placement] for you as well?
Yes I mean, yes you socialise and you interact with the public. You know, the elderly, they have got a few stories to tell so it's interesting.

So do they take [an] interest in you?
Well I should think so, yes, I try not to be too dull. (LP P, Aged 25–34, Mixed Ethnicity)

This example demonstrates how working with older women constituted a safe space for this LP to practice his non-prison-focused social skills. However, gendered relations can be a potential challenge for both LPs and placement managers, especially where prisoners' movements between an almost exclusively male prison environment and the highly feminised charity retail sector has prompted negative reactions, e.g., with LPs who have resented female shop managers managing them and their work.

Other LPs stressed how they valued the opportunity to present themselves as 'normal' and shifting perceptions of their identity when working with non-prison colleagues in the locale of the charity shop:

Interacting with people, getting to talk to people to be honest, that was the best thing of all. People I thought I wouldn't get on with it but I did. (LP M, Aged 25–34, Mixed Ethnicity)

> I suppose it gives them a chance to see that like we are normal people, we are not just like, obviously the big picture of what criminals are like, we are normal people, everyone has broken the law, just some people got caught for it. Obviously there are bigger issues behind it, yes, people do have a … they do see criminals as different people but we are no different from anyone else really. I think that is good … seeing that we are trying to change our lives around. (LP J, Aged 35–44, White British)

A different form of social mobility can be identified in those former 'white collar' LPs for whom their Community Service placement – rather than the experience of being in prison – provided insight to the lives of those living on low incomes. When asked what he had gained from his charity shop placement, one LP, convicted for fraud, identified the social deprivation of the community where the shop was located as even more 'eye-opening' than being in prison, and he saw this as one of the key benefits of his experience:

> I get out of the prison every day which is good, the shop was in quite a deprived area and I think it was quite an eye opener for me even working in the shop, this is a big eye opener for me being in prison, but working in a charity shop and sort of seeing … all the charity shops, betting shops and pound shops there, and it is really a way of life for the people living there, you see the same people every day in the shop and it is just very different from anything I have been used to … it gave a bit of experience, a bit of perspective on to a very different way of life almost … I think it is quite interesting because I have always worked in financial institutions, worked in my own businesses, I have never worked in a so-called blue collar environment.

He went on to highlight the value of transitioning back to the world of work through Community Service placements:

> I think it was quite interesting for me, I actually enjoyed it and it was quite interesting for myself to actually see that, well are you going to go in there to work and I did, I have worked really hard, I have done everything from climbing up ladders, you name it, to doing all the various roles in there. I think it is a good introduction back into the world of work. (LP E, Aged 35–44, White British)

Given the inverse relationship between employment and reoffending rates after release, preparation of LPs for post-release paid work is a key motivation underpinning prison authority adoption of prisoner Community Service under license. Ten per cent of respondents interviewed had never experienced formal employment, making the 'work experience' element of Community Service particularly formative. LP P, who had left school at age 15 and was serving a sentence for robbery, reported the combined benefits of social interaction and work experience in his Community Service placement. This provided a slow and gradual transitional process (see Cresswell 2011, 2012; Fullagar et al., 2012), towards release and being 'release-ready':

> You feel positive when you are out there because you know you are getting closer and you are interacting again. Because you do lose it, when you haven't had any interaction for so long and then to get thrown back into it … Some people must lose their ability to communicate with people, it's hard to explain. They just lose that ability to just be able to communicate. They lose self-confidence, I think a lot of people [do] and obviously …. that's why D Cat's [Category D prison is] good you see, because it puts you back in and it reins you in slowly, slowly, and then eventually you get released. So I think that's why it's good … it's got to help because it's slowly, slowly, weaning you back in … like interacting with people, your customer skills obviously, you go into work again you are going to need to bring up your skills obviously if you got a bit rusty over time you've been away. Books, if you do some sort of books you know, anything, helping customers… (LP P, Aged 25–34, Mixed Ethnicity)

A minority of LPs interviewed experienced Community Service work as judgemental and exploitative. LP V, who spoke highly of his experience in the advice centre placement, reported a very different experience when placed in a charity shop for three weeks, where he felt both the work and attitudes towards him were inappropriate:

> The atmosphere, it was very, you tread on egg shells. You felt intimidated, you felt that you were just there to be someone's skivvy, you know, you didn't have a say in anything, I mean one of the questions was 'can you make me a cup of tea?' and my answer was no because I don't drink tea so I don't make tea, I am not here to make you tea, I'm sorry but if you are going to treat me like on the basis that I'm here to work then I will treat you back the same way. It is a two way thing … because I have had businesses on my own and I know how to treat staff, it wasn't inviting, it wasn't okay. At the end of the day they don't know what background you have come from. You don't know if the guy has done a couple of months in prison or 10 years in jail and it is like you are treating me the same, 'if your face fits' sort of thing. I hated the three weeks I was there, I absolutely hated it but I went along because if I didn't go I will only get myself into more trouble, so I did it because I had to do it. (LP V, Aged 35–55, British Asian)

In this case the LP experienced very little social mobility within the charity shop because of a mismatch of placement and perceived negative attitudes of some shop staff. Managers play an important role in establishing the ethos of the charity shop environment (Maddrell, 2000), but the ambiance of the often-confined work space and LPs' sense of social ease and their subsequent social mobilities also depend upon the people they are working alongside, as one LP articulated:

> I didn't enjoy nothing, I didn't gain anything … It was an absolute waste of time … it doesn't help anything, getting any kind of skills, I mean places you

go to work are [should be] welcoming, they don't treat you like a working person or a colleague, there is that distance like you are the prisoner and they are the staff, and they made sure you know that … I am an educated guy, I am not one of them guys that are looking for an easy ride, I do my graft, whatever it is I've got to do … The manager, a couple of paid staff, some volunteers were fine, not everyone is the same, some were great, but the majority there, you know, you knew where you stood. (LP V, Aged 35–44, British Asian)

This highlights the importance of matching LP skills and interests to the placement where options are available; however, prison officers reported having insufficient time within their multiple work roles to always be confident of providing optimal placements and preparatory briefings for LPs, including explaining the personal and wider value of their community service (Interview, Prison Officers (PO) 1, Male, White British).

By contrast, to LP V, the majority of LPs reported the benefits of their placement as preparation for paid work and release, particularly when compared with the inertia and immobility associated with much of prison life:

I think [charity X] are fairly renowned for making you work fairly hard, but I don't mind that at all, I think it is very good preparation, because the danger is you are lying on your bed potentially, that's what the vast majority of people do because there isn't a lot else to do, … in some cases years and years and years, and it is a nice almost sort of half-way house between being in prison and doing nothing and going back out and hopefully getting a job and going back into the world of work. (LP H, Aged 45–54, White British)

Here, the LP's use of the metaphor of 'halfway house' underscores the embodied and experiential qualities of Community Service as transitional time-space.

Maximising the benefits of Licensed Prisoner transitional mobilities through charity shop placements

There are clear potential benefits for LPs undertaking Community Service beyond it being a necessary and compulsory stage towards being permitted to apply for paid work under the same licensing scheme, and ultimately parole and release – the ultimate mobility from a prisoner perspective. The key mobilities experienced are physical, emotional-psychological, and in terms of work experience and future employability. More specifically, Licensed Prisoners undertaking Community Service experience mobilities through (a) leaving the confines, milieu and regime of the prison; (b) travelling to and from the prison; (c) experiencing social interactions and roles which are not defined by being a prisoner or prisoner hierarchies; (d) being 'transported' emotionally and developing enhanced self-esteem through work is both appreciated and gives satisfaction through a job well done; (e) providing a new perspective of LP positionality in relation to the socio-economic deprivation or life challenges faced by others, e.g., local or international poverty, or those living with disability or life-shortening illnesses; (f) undertaking regular working hours, which may be experienced as re-habituation or as a novel experience, which in turn impacts on attitudes as well as embodied experience, including stamina and body clock; (g) attaining or (re)applying workplace skills and experience; (h) gaining a sense of well-being (and possibly reparation) through doing something which benefits others; and (i) gaining curriculum vitae content and a reference.

Occasionally these benefits culminated in a further mobility in the form of a post-release job offer for one of the paid posts within a charity retail chain, as was the case for one LP who was unlikely to be re-employed in the banking sector but who brought many transferable skills from his former profession and had proved himself to be highly effective across a number of roles during his Community Service placement. However, he did not wish his particular journey to charity shop management to be publicly known, especially by the current LPs undertaking their own Community Service in the shop he now managed. For him, his Community Service had facilitated his post-release employment, thereby mobilising his return to paid professional work, clearly central to his personal identity, including putting his custodial sentence behind him – another form of 'forgetting'.

However, not all opportunities to develop employability are being capitalised upon within the placement scheme. Some charity retail operations offer volunteers an opportunity to complete a Non-Vocational Qualification (NVQ) in Retail skills, but, as a manager explained: '[W]e haven't had anyone take it up from the prison or the [Probationary] Community Service .... Quite often the problem is with CS [Community Service] or the prison [volunteers], they are not here long enough to complete the course' (Charity Shop Manager 1). A nationally recognised qualification such as an NVQ would provide tangible proof of skills and training, aiding LP movement into employment after release, but facilitating LP access to these qualifications while on CS placements requires both 'joined up thinking' and aligned time frames on the part of prison authorities and charity retailers.

Reflections on carceral mobilities

In light of the range of temporary mobilities experienced through time spent outside of the prison on licensed Community Service placements (as part of a wider pre-release programme including paid work and home visits), it is necessary to shift thinking about carceral spaces from Goffman's notion of the prison as 'total institution' to 'understanding of confinement as a dynamic and often contradictory state of betweenness' (Baer and Ravneberg, 2008: 205). As with other states of 'betweenness' such as bereavement or pilgrimage (see Maddrell, 2013; Maddrell and della Dora, 2013, respectively), transitional liminal qualities can characterise this experience. This porosity of prisoner spatio-temporal experiences has been identified in relation to a variety of contexts such as Community Service (e.g., Horne and Maddrell, 2002), home visits (Baumer et al., 2009), Electronic Monitoring (Gill, 2013), and *etap* (Moran et al., 2013b), requiring research which is attentive to the complexities of mobile, embodied, and transformative carceral contexts (Moran et al., 2013a). Within this context, the embodied and emotional transitional-liminal time-space of Community Service placements merits further analysis. The moorings or tethers of prisoners' transitional mobilities are writ large in the frustration of the enforced inertia of a punitive 'lay down' or waiting for administrative processes to be completed, e.g., an interim between a Community Service Placement and day release for paid work.

These intermeshed complexities were highlighted by LP E: 'I have been in here [prison] 2 days, 3 days since the end of my community work and I have hated every one of those three days'. LP P (Aged 25–34, Mixed Ethnicity) stressed that while it was good to be out of prison, there were clearly defined constraints on his movements, notably being allowed out of the shop only during his lunch break. This contrasts with the initial cohort of LP interviews in 1998–2000, when some trusted prisoners helped collect bags of donated goods from residential areas and ran errands outside of the shop. One of the shop managers who was an early adopter of the use of LP Community Service explained that she liked to take an LP with her when collecting donated goods because of his physical strength and because she had learned that long-term inmates can lose their long-distance vision due to the limited lines of sight experienced in confined spaces (Manager Q, Oxford). Wider professionalisation of the charity retail sector, including out-sourcing of door-to-door collections of donated goods, will have reduced such opportunities for LPs across the time frame of this longitudinal study, but shop managers, prison officers, and inmates all referred to the current strict protocol for LPs on Community Service placements, including no contact with people known to them, and nothing to be taken in or out of the prison, including money:

> When they sign their license they also sign their community work contracts, which is a breakdown of what they can and can't do. They sign it, so if they do make a mistake they can't turn around and say 'Well we didn't know that': 'Yes you did because you signed your contract' ... So no visiting licensed premises, do not meet family and friends, no mobile phones, the usual sort of thing. (PO 1, Male, White)

This regulatory regime formalises expectations and constraints on LP behaviour and opportunities; over the last 10 years additional regulations have been insti-tuted in response to actual or potential breaches of security, e.g., absconding, theft, or acquiring drugs, the consequence being greater control of physical mobility for LPs on Community Service. These rules also preclude informal contact with friends and family and opportunities to purchase clothing, books, and electric goods while at the charity shop, each of which represented a form of social mobility for LPs.

It is the individual prisoner who is licensed rather than space of charity shop, and keeping to these regulations is ultimately the responsibility of the LP and a matter of self-regulation and discipline, but the staff of the shop, especially shop managers, have a duty of care to both the Prison Service and prisoner volunteer, as well as to other volunteers, shoppers, and the charity itself. Shop managers were often keen for prisoners to know they were strict in applying the contracted regulations and would not hesitate to report significant infringements, whereby the LP would be physically returned to prison, privileges revoked, and the progress towards parole halted. Whilst managers may have little choice in such a scenario, it highlights both the regulatory role they play in the lives of the LP and the conditionality of placements, which require both self-regulation and fulfilment of the placement manager's expectations in order to fulfil the license contract. It is also significant that access to placement Community Service or paid work can be withdrawn as a result of any breach of conduct within the boundaries of the prison itself as well. This underscores the risk as well as opportunities constituted through placements, whereby the transition to parole and release may be jeopardised as much as facilitated by Community Service. Ultimately the mobilities offered by such schemes can be withdrawn and thus represent the fragility of the mobile relations of LPs.

Hence, Prison Officers discouraged allowing LPs to work on the till during charity shop placements (PO 3, Male, White). This position was echoed by several LPs who saw any involvement with money to be risky at a stage in the lifecycle of their sentence when any question of misconduct might cost them their placement and subsequent access to paid work and, ultimately, parole:

> No, I didn't want to have to do that [the till], involving money, I thought it would save me the hassle if anything went wrong so they couldn't accuse me of anything ... because I have heard stories ... if money goes missing you are going to be the first one to blame, so to save that happening, they asked me if I wanted to go on the till, but it just saves a lot of hassle really, to be honest. (LP M, Aged 25–34, Mixed Ethnicity)

Prisoners who have gained privileges and/or are near to release have a lot to lose and therefore evaluate and manage their own risks in order to protect their mobile trajectory towards release. A placement may reject an LP as unsuitable for the post, or may have them recalled for breach of licence and/or conditions of the placement. Similarly, charity shops and other Community Service placements may be dependent upon the labour provided by LPs but find the prisoners are withdrawn or transferred, or the prison has a 'lock down', these LP mobilities and immobilities each resulting in unpredictable staff shortages for CS placements.

Conclusion

Prisoners' mobilities within the context of Community Service placements are highly conditional, vulnerable, and vary enormously. Interview material discussed above highlights a number of significant transitional time-space processes experienced through LP Community Service placements in charity shops. Mutual social acceptance has been shown to be pivotal, albeit this may be something achieved over time rather than immediately; the use of existing skills and/or training for new workplace skills is also important; each contributes to the success of CS placements, including LP social ease, associated micro-mobilities, and subsequent employability. Placements can represent either temporary experience of the world outside the prison sector for those on indeterminate sentences, as well as an important and constructive stepping stone on the longer transitional journey towards release and legitimate occupation/employment. However, the full range of potential mobilities and their associated benefits that may be accessed through Community Service placements in charity shops are experienced unevenly. Social and spatial practices continue to be controlled within defined boundaries, that is, a form of virtual confinement, established by the contract for day release under license but enforced primarily through self-regulation. Thus charity shops are fulfilling shadow state roles in supervising and rehabilitating LPs, and this may be expressed and experienced as Panopticon-like control; but by and large, the risk to the transitional mobilities associated with the lead up to, and ultimate release, cause most LPs on Community Service placements to modify and limit their immediate behaviour and mobilities in the interest of the end-goal of release. While many experience Community Service placements in charity shops and other community projects as a mediated form of liberty, self-expression, opportunity for socialisation, and skills practice or development, such opportunities are undermined by ill-judged or poorly briefed placements and infelicitous attitudes on the part of the LP or host institution. Notably, with a more flexible placement programme HMPS could work with charity retailers offering workplace qualification training opportunities in order to maximise benefits to selected LPs. Likewise, greater time availability for POs to brief LPs on the personal and wider benefits of their placements would help motivate LPs to respond to and benefit from the work environment of the placement, which in the first instance may be experienced as a rude awakening rather than gradual transition towards being release-ready.

LP Community Service placements in charity shops contribute to our understanding of the dynamic assemblage of permeable spaces and boundaries that can constitute carceral spaces and practices in the UK and the transitional mobilities associated with the temporal life cycle of a sentence. The conceptual frame of mobilities, incorporating movement, travel, meaning, and embodied-emotional experience, highlights Licensed Prisoners' liminal status as both prisoners and 'volunteers', as well as their varied negotiations of the equally liminal and transitional time-space constituted by Community Service, e.g., moving between and operating in the different social and emotional 'worlds' of the known prison regime and the unknown charity shop placement. After reiterative prison routines, day-release Community Service placements may be experienced as a process of gradual transition towards release or dizzying oscillation between prison and non-prison social worlds. This understanding further underpins a critical evaluation of the charity shop as an unproblematic space of prisoner reparation and the gradual staged movement towards parole or release and rehabilitation into wider society. Furthermore, the very *plasticity* of carceral regimes and spaces discussed here indicates the need for a conceptual frame which can incorporate a *dynamic carceral assemblage* of relational spaces and practices, including, but not limited to, physical and regulatory fixed point nodes/moorings; bounded and (semi)permeable spaces; physical and imaginative journeys; and transitional personal, social, and professional mobilities, which may be experienced as gradual or staged.

Acknowledgements

My thanks go to the prisoners, prison officers, and charity shop managers who shared their experiences and insights with me. Thanks also to the editors of this volume for their insight and patience.

Note

1 Community Service Orders may be issued by courts in lieu of a custodial sentence and are typically measured in hours; e.g., an offender may be put on Probation and required to complete 100 hours of Community Service; or, as the main focus of this study, offenders serving custodial sentences may be given the opportunity to undertake Community Service activities within or outside the prison, subject to review and risk assessment. Prisoners in 'open' prisons commonly undertake Community Service. Open prisons house low-risk prisoners; they have permeable boundaries in contrast to 'secure' prisons and allow approved prisoners regulated home visits and other sanctioned temporary exits from the prison campus.

References

Baer LD and Ravneberg B (2008) The outside and inside in Norwegian and English prisons. *Geografiska Annaler. Series B, Human Geography* 90(2): 205–216.
Baumer EP, O'Donnell I and Hughes N (2009) The porous prison: A note on the rehabilitative potential of visits home. *The Prison Journal* 89: 119–126.
Cresswell T (2011) Mobilities I: Catching Up. *Progress in Human Geography* 35(4): 550–558.
Cresswell T (2012) Mobilities II: Still. *Progress in Human Geography* 36(5): 645–653.
Cresswell T and Merriman P (2011) Introduction. In: Cresswell T and Merriman P (eds) *Geographies of Mobilities: Practices, Spaces, Subjects*. Farnham: Ashgate, 1–18.
Fullagar S, Markwell K and Wilson E (2012) (eds) *Slow Tourism: Experiences and Mobilities*. Bristol: Channel View Publications.
Gale T, Maddrell A and Terry A (2015) Introducing sacred mobilities: Journeys of belief and belonging. In: Maddrell A, Gale T and Terry A (eds) *Sacred Mobilities: Journeys of Belief and Belonging*. Farnham: Ashgate, 1–16.
Gill N (2013) Mobility versus liberty? The punitive uses of movement within the outside carceral environments. In: Moran D, Gill N and Conlon D (eds) *Carceral Spaces: Mobility and Agency in Imprisonment and Migrant Detention*. Farnham: Ashgate, 19–35.
Hannam K, Sheller M and Urry J (2006) Mobilities, immobilities and moorings. *Mobilities* 1(1): 1–22.
Horne S and Maddrell A (2000) Editorial: Charity trading. *Journal of Non-Profit and Voluntary Sector Marketing* 5(2): 101–102.
Horne S and Maddrell A (2002) *Charity Shops: Retailing, Consumption and Society*. Abingdon: Routledge.
Levinson J and Farrant F (2002) Unlocking potential: Active citizenship and volunteering by prisoners. *Probation Journal* 49(3): 195–204.

Maddrell A (2000) 'You just can't get the staff these days': The challenges and opportunities of working with volunteers in the charity shop, an Oxford case study. *Journal of Non-Profit and Voluntary Sector Marketing* 5(2): 125–140.

Maddrell A (2011) 'Praying the Keeills': Rhythm, meaning and experience on pilgrimage journeys in the Isle of Man. *Landabrefið* 25: 15–29.

Maddrell A (2013) Moving and being moved: More-than-walking-and-talking on pilgrimage walks in the Manx landscape. *Journal of Culture and Religion* 14(1): 63–77.

Maddrell A and della Dora V (2013) Crossing surfaces in search of the Holy: Landscape and liminality in contemporary Christian pilgrimage. *Environment and Planning A* 45(5): 1104–1126.

Michalon B (2013) Mobility and power in detention: The management of internal movement and governmental mobility in Romania. In: Moran D, Gill N and Conlon D (eds) *Carceral Spaces: Mobility and Agency in Imprisonment and Migrant Detention.* Farnham: Ashgate, 37–56.

Ministry of Justice (2012) *Compendium of Reoffending Statistics and Analysis 2012.* Available at: www.gov.uk/government/statistics/compendium-of-reoffending-statistics-and-analysis

Moran D, Gill N and Conlon D (2013a) Introduction. In: Moran D, Gill N and Conlon D (eds) *Carceral Spaces: Mobility and Agency in Imprisonment and Migrant Detention.* Farnham: Ashgate, 1–9.

Moran D, Piacentini L and Pallot J (2013b) Liminal transcarceral space: Prison transportation for women in the Russian Federation. In: Moran D, Gill N and Conlon D (eds) *Carceral Spaces: Mobility and Agency in Imprisonment and Migrant Detention.* Farnham: Ashgate, 109–124.

Prison Blog Spot (2015) *Going on a Townie: Day Release from Cat-D.* Available at: http://prisonuk.blogspot.co.uk/2015/01/going-on-townie-day-release-from-cat-d.html

Rose N (2000) Government and control. *British Journal of Criminology* 40(2): 321–339.

Turner V and Turner E (1978) *Image and Pilgrimage in Christian Culture: Anthropological Perspectives.* New York, NY: Columbia University.

Turner J (2016) *The Prison Boundary: Between Society and Carceral Space.* London: Palgrave Macmillan.

Winnicott DW (1971/2005) *Playing and Reality.* London: Routledge.

17 Prison

Legitimacy through mobility?

Christophe Mincke

Although prison as punishment is as old as modern liberal and democratic states, from the beginning it has faced continual challenges over its legitimacy. Human rights and prison were conceived at the same epoch (during the eighteenth century), but have always been in tension. Accordingly, thinkers and policy makers have constantly had to re-examine and consolidate the legitimacy of the latter. This question of legitimacy was logically grounded: the prison was meant to deprive of what was considered as essential in society. The punitive purpose could not be served by the deprivation of second-order interests such as social or religious status or reputation, even though deprivation of first-grade rights was hard to legitimate. This presented a carceral paradox.

The fact that the prison in the nineteenth century proved to be inefficient and, worse, counter-productive, lent it even less legitimacy. Not only was its legitimacy dubious, its efficiency was also highly controversial. At the end of this century, the urge for a profound reform of reactions to deviance resulted in the development of 'rehabilitative' theories, grounded on medical and scientific approaches and treatments of persons diagnosed as dangerous. But the dream of replacing the prison by a medico-social system never became reality. And now at the start of the twenty-first century, the 'moribund' prison of the late nineteenth century is still alive and well. The medico-social system was simply added to it, causing some evolution, yet the core of the project has remained unchanged. And the question of legitimacy is as relevant as ever. Now, the prison must face its critics. Too expansive, inefficient, inhuman, deprived of any positive project for the detainees, it has to be reinvented once again or else disappear.

One of the characteristics of an institution such as the prison is its capacity to endorse very diverse projects and legitimising discourses. It remains that the core of the carceral project is to detain and *immobilise* people behind walls as a form of retaliation for the crimes they committed. This can be considered as the base line of modern imprisonment and will be the starting point of this chapter. I shall ask how the painful immobilisation at the base of the prison could legitimate it today – or not – and show that a new legitimation discourse is emerging, which introduces a new carceral project, based on mobility as a value in itself.

The Prison Act

To achieve this goal, I shall rely on the theory of mobilitarian ideology developed over recent years (Mincke, 2013, 2014a, 2016; Mincke and Lemonne, 2014) and on empirical material: the preparatory documents of the recent Belgian Prison Act (2005).[1] The Prison Act – the first in Belgian history – aims at regulating the inner functioning of the prison in light of the purpose of the punishment, control of its execution, communication with the outside, visits, complaints and disciplinary procedures, security regime, and so on. The parliamentary documents, mainly in the report of the 'Dupont Commission', which prepared the project, contain a lengthy consideration on prison, its purposes and its ideal functioning, providing a 'serious discourse' (in Foucault's sense) (Dreyfus et al., 1984: 76–77) on prison and its legitimisations. The parliamentary process was the occasion for a broad reflection about the prison of today and tomorrow.

In what follows, I shall first briefly review the Prison Act which I am about to analyse here. Then, in a second section, I shall develop the classical vision of imprisonment as a painful immobilisation. Thirdly, I shall present a new conception of the prison challenge: its effect on mobility and autonomy. Finally, I shall provide a new interpretation of prison, based on mobility tests, rather than painful immobilisation.

Painful immobilisation of classical prison

The classical (modern) conception of the prison is grounded in the idea of painful immobilisation. The aim is, first of all, to punish by inflicting a pain in retaliation for the offence. The fact that a prohibition has been violated is justification for the state's punitive reaction. In the context of modern democracies of the late eighteenth and the nineteenth century, in which freedom was a central value, it made sense to conceive of punishment as being deprived of it. The prison was thus an immobilisation *dispositif*, in many ways.

Immobilisation

First of all, deprivation of liberty is a particular punishment based on physical immobilisation. The liberty to circulate is suppressed. The aim of the institution is thus to literally prevent incarcerated individuals from moving. Yet the institution is not just a form of organisation. It is housed in a specific building, conceived as an immobilisation tool. The purpose was to cut the inmates off from the free society and vice versa (Demonchy, 2004: 282–285). On the inside, the prison is compartmentalised: the prisoners are not allowed to move freely. In the star-shaped prison, the inmates stay in cells for most of the day, while the guards move. The movement of the guards is at the base of their power (Bauman, 2000: 9–10), unless they in turn are under the surveillance of the central control centre, with the Benthamian Panopticon applied to them (Demonchy, 2004: 284–287).

Secondly, the prison's social space bears the traces of the compartmentalisation of the physical space. Inmates are (theoretically) divided into categories (Maes, 2009: 229–276) in order to prevent contacts and contamination. In the same way, the institution's different functions are rigidly distinct: inmates, guards, administration, chaplain, families. To the physical multilevel segregation, corresponds a social segregation. The prison aims to maintain the inmates in a space separated physically and socially and to prevent them from moving from the place they have been assigned to.

Thirdly, immobilisation also consists of preventing prisoners from acting as normal persons, citizens, and legal subjects. They are deprived of the rights that make it possible to be part of the political and juridical community named the 'state'. They cannot freely come into contact with others. Their physical immobilisation is coupled with a social and juridical one. Fourthly, immobility is not exclusively about spaces, but also about time. On the one hand painful immobilisation also becomes reality through the repetition of an ever-recurring routine during a well-defined stretch of time. On the other hand, time in prison was purely repetitive, each day the same, with nothing expected to happen. Even time stood still. Accordingly, the classical prison is a total institution aiming at a multidimensional immobilisation, in physical and non-physical spaces.

The reign of boundaries

Immobilisation makes sense in the context of a society that prizes anchorages: family, nation, culture, class, country, or even the company, with every individual asked to stabilise in different spaces. In the context of a sedentarist metaphysics, roots are part of human experience: 'to be human is to have and to know your place' (Cresswell, 2006: 31). Freedom was then thought as the ability to move from one anchorage to another (Bauman, 2000: 32), respecting the rules, the first of which is the rational decision to move only if it is needed (Cresswell, 2006: 29). Nomadism, cosmopolitanism, internationalism, or migrants were seen as potential dangers, resulting in measures to control and anchor the concerned populations. Nomads, Jews, tramps, moving workers, or traveling political opponents were given special attention (Cresswell, 2006: 26). Discipline and social functioning were gained through the assignment of a geographical, social, professional, and familial position. Anchorage was conceived as necessary, and mobility, as important as it may be, was only a secondary phenomenon. In such a context it seems logical for ill-integrated persons and offenders to be disciplined through immobilisation, as was the case in Belgium, where vagabonds could be incarcerated for being homeless,[2] or where workers could travel only if they could prove they were moving from a factory to another, where they had an employment (Delwit and Gotovitch, 1996). Compulsory rooting was a way to prevent disorders inevitably arising from uncontrolled mobility. Thus, to understand the immobilising prison, one must question the value socially attributed to space, time, and (im)mobility. The legitimacy of prison as punitive immobilisation depends on the social representation of space, time, and mobility. Logically, this vision of prison comes about in a context that it can echo.

Indeed, the classical prison is intimately linked to a particular conception of space-time that we call 'limit-form'. In this spatio-temporal morphology, space is seen as unorganised until it has been circumscribed by a boundary. Nation-states, political authorities, or private properties are conceived through the category of the boundary. This concerns both physical and social spaces. The scientific disciplines, social classes, genders, functions in an organisation, families, and the competencies of authorities all are structured through sets of boundaries. Each territory can itself be divided in sub-territories: the nation-state can be divided into provinces or departments, legal jurisdiction can be attributed to the central state or to administrative subdivisions, and scientific disciplines can be divided in more specialised sub-disciplines (Montulet, 2005: 148). The closure of the *prison*, its inner compartmentalisation, classification of inmates, or the clear distinction of functions inside the institution are some of the physical and non-physical boundaries that constitute the space of the classical prison. The whole carceral system is conceived and organised through such circumscriptions.

But a boundary cannot exist without permanency (Montulet, 2005: 145). Its ability to structure the space over the long term makes it a 'proper' boundary, rather than a purely arbitrary and versatile subdivision. What would be the point of a limit between states, families, or sexes if at any moment one could chose to freely cross the border or modify its outline? Thus, these boundaries are necessarily linked to a specific conception of time as a succession of stasis and ruptures. The national border remains unchanged until, brutally, a treaty or a war modifies it. One is situated in the organisation chart of one's company until, suddenly (and maybe expectedly), one's position changes due to a promotion. A man is an inmate until the day he is released. Social organisation, morals, management, legal rules, and so on command the definition of fixed borders that will be strictly kept until the pressure for a change is high enough. Then a brutal adaptation occurs that leads to a new stable position.

In this system, the relation to the carceral system is not made of constant evolution (variations of the control level, scalable regimes, semi-detention, release on parole, and so on). It is a question of entering the prison space until the precise day of liberation. Transitions are pure ruptures: loss of liberty and recovery of it. Thus the space-time morphology we call limit-form is this combination of a space structured through boundaries and a time of stasis and ruptures. In this context, entities are assigned to specific and well-defined spaces. Anchorage is thus prior to mobility: social class, social statute, gender, function in a company, scientific discipline, nationality, language, or religion define the territories anyone belongs to. Mobility is possible, but it takes the form of a *crossing* mobility. Boundaries can be crossed under certain conditions in order to move from one anchorage to another.

Relocating the prisoner

This framework reflects the way the prison articulates with the free society, with walls and decisive moments (sentencing, arrest, release, strictly limited visits) and a mobility (between the inner and the outer, between the category of free citizens and the one of condemned) through boundary crossings. Thus, the prisoner is removed from society, anchored in jail, and then, at his release, expected to (re)anchor in the society and in the (appropriate) social groups. In prison, the inmate is required either to stay inactive in his cell or to develop a useful stereotyped activity in the case of forced labour or else therapeutic interventions. The good prisoner, like the good citizen or worker, is a well-*functioning* entity in a mechanically conceived institution. Therefore, the inmate is expected to comply with normative and curative injunctions that the prison imposes to discipline him. *Application* of orders is imperative, regardless of any personal initiative or creativity. Nothing is asked of the prisoner other than discipline (Mincke and Lemonne, 2014: 4–6).

On the other hand, the inmate is part of a population, managed as such by the institution in a Foucauldian biopolitical approach. *Incorporation* into categories (convicts, subcategories of inmates, mentally ill offenders, and so on) is the key for the system to manage the prison population. The 'good' inmate is also the one who complies with the requirements and rules of the group he belongs to, and whose destiny follows the statistical previsions.

Stabilisation, then, is the aim, whether it is in prison (for a stretch of time), as part of a certain category or, back in the free society, in a socially useful function, group, or job role. Again, the question involves where the individual belongs and how to make him stay in the right place. To (re)embed (Bauman, 2000: 32–33) individuals who, through delinquency and socially disordered behaviours, have demonstrated disrespect of social and legal borders is thus one of the key aims of prison. The classical prison is an anchorage-based prison, based on clear boundaries and a scansion-based time. *Functioning, application, incorporation,* and *stabilisation* are its key principles. This is what underlies the prison of painful immobilisation.

Sketches of a new prison

According to what has been written above, the prison was not a strange exception but a radical application of the common representational categories of space-time and mobility: the limit-form and crossing mobility. But is this view of prison still up to date? Examining the parliamentary documents of the Prison Act, in search of traces of a new conception of prison, this chapter questions this basis for the legitimacy of the prison.

Dependence and openness

First of all, the documents contain vast considerations about the reason why prison is problematic (and requires a Prison Act). This helps to define the core of the legitimation problem of prison. Is the 'issue' with prison linked to the fact that it deprives people of a fundamental right, freedom of movement, necessitating a special attention and effort to limit the possible excesses? Not really. Rather, as Decroly and Van Parys note,

Regulation and supervision by the penitentiary places the inmate in a situation of heavy dependence on others. This stunts all his moral resistance at the social level and the flexibility that enables him to function in a normal context, conditions that are necessary in the future if he is to continue living in respect of the law in free society.[3] (Decroly and Van Parys, 2001: 66–67)

It is thus a prison-induced *dependency* that justifies a special concern about imprisonment and, in the end, the adoption of the Prison Act.

The Prison Act therefore pursues a main goal:

Preventing or limiting the adverse effects of confinement … implies the suppression as far as possible of the prison as a 'total institution', the maximal normalization of daily life in the prison, an opening as broad as possible to the outside world and the definition of a carceral trajectory placed in the perspective of early release. (Decroly and Van Parys, 2001: 69)

If normalisation is the goal, the prison cannot be conceived as hermetically closed (physical and non-physical) spaces. On the contrary, it is meant to be as open as can be,

both from the perspective of limiting the detrimental effects and in granting the detainee the possibility to take up responsibilities (e.g., as a parent or spouse), and in the context of reintegration, the existing provisions in terms of visits have been significantly expanded. (Decroly and Van Parys, 2001: 137)

As such, '[t]he principle of equivalence is strongly linked to the continuity principle … according to which the inmate is entitled during his prison term to a continuation of health care similar to that prior to his imprisonment' (Decroly and Van Parys, 2001: 167). The prison becomes more than just the space between the walls, but a widespread phenomenon. It must be an open space, where passers-by come and go.

Autonomy

Inside the prison, the standard regime has become community living. The aim is to help creating a real social space inside the prison, instead of trying to cut inmates from their relationships.

The ordinary Community regime allows inmates to spend their detention time in community living and working areas and to participate jointly in organized activities in the prison (art. 49) … the stay in the individual living space is not considered as a form of exclusion from the community of prisoners, but as an opportunity to exercise the right to privacy. (Decroly and Van Parys, 2001: 128)

Furthermore, the prison can no longer be an institution that seizes the individuals and forces them to comply with pre-established statutes and roles. It becomes a participatory project.

It is possible to avoid to a very large extent the harmful effects of detention if the problem of the form to give to the execution of the prison sentence is not addressed primarily from the penitentiary institution and its interests, but also from the world of convicts themselves, values and interests they consider worthy of being taken into consideration, as well as from the representation they have of their necessities and their needs … It is therefore appropriate … to consider the convicts as valid interlocutors and full partners when it comes to dialogue. (Decroly and Van Parys, 2001: 70)

In this perspective, when it comes to participation, the rules are the same for everyone.

Different principles contained in the draft law, namely the principles of respect, normalization and participation, shall apply by analogy to staff. (Decroly and Van Parys, 2001: 125)

In this context, there can be no more pure and simple support from the administration and socio-medical experts. Help can only be a proposal made to the inmate, whose duty it is to identify, select, and participate in activities that can be useful to him.

All effort must be expended during the time in prison time in order to … make available to the inmate an offer – with no imperative nature – of activities and services as varied as possible, corresponding as closely as possible to his necessities and needs, particularly in view of his future reintegration into free society. (Decroly and Van Parys, 2001: 74)

The participatory logic implies that there is no obligation for a consensus on the meaning of an action as long as participants agree with the action itself.[4] In this sense, prison is said to have no goal in itself, but to rely on the prisoner himself: 'the convicted person is responsible for the meaning to be given to the detention because, after all, this is "his" sentence' (Erdman et al., 2003: 49).

But, if prison has no aim in itself, particular aims can be pursued during the incarceration.

Formulating such objectives must therefore necessarily be preceded by or be inextricably linked to the priority objective, which is to limit the adverse effects of detention and that, as a necessary condition for the implementation of constructive goals and in order that convicts collaborate to their achievement. (Decroly and Van Parys, 2001: 65)

Therefore, people are put in prison without predetermined objective, but objectives are pursued, the first of which is the limitation of the detrimental effects of the treatment inflicted to the detainees. This should logically lead to the suppression of prison, as it is the only way to truly achieve this goal. Everything that makes prison a special place should be abolished: geographical closure, exclusivity to serve prison penalty, unusual life conditions, limitation of movements, restriction to social contacts, and so on. In any case, prison-time should no longer be a lost stretch of time. The Prison Act does not rely on the idea that prison punishment is perfect through pure immobilisation, without anything to do or to hope for. Incarceration opens a time of projects and activities.

> [R]ehabilitation must allow the prisoner and free society to consider the detention time in terms of repaying a debt, a significant contribution to reinstating the person concerned in his honour and rights. (Decroly and Van Parys, 2001: 82)

> Regarding convicts, whether as completion of an unfinished education, retraining, refresher courses, vocational training or continuing education, this right is then correlated with an individual detention plan, based on which the needs of the prisoner will be established, in consultation with him. (Decroly and Van Parys, 2001: 147)

> [I]n the semi-detention regime ... the convicted person is given the opportunity ... to systematically leave the jail during the time required to: 1° engage in activities whose continuation is likely to prevent his retrogression, such as fulfilling household tasks, continuing vocational training, internships or study; or 2° provide restorative services towards the victims of the offenses for which he was convicted. (Decroly and Van Parys, 2001: 383)

What is demanded of the inmate, first of all, is to preserve or restore his autonomy and capacity for action and adaptation. In return, imprisonment can no longer be a standard punishment, foregrounded on processes of immobilisation, but has to be adapted to the trajectory of each individual:

> [T]he individual detention plan ... can be considered as an instrument to individualize the punishment of deprivation of liberty, as a necessary condition to humanize it, there must be opportunities for differentiation in the planning and in the form of the prison stay, where a staggered detention should be in the context of a gradual increase in the freedom to come and go. (Decroly and Van Parys, 2001: 418)

Though there remain evident traces of classical imprisonment in the preparatory documents of the Prison Act, it is clear that it largely promotes a very different vision of the prison. Imprisonment takes place in a (socially and physically) open and decompartmentalised space, to the point that the prison building loses its monopoly on the implementation of prison sentence. Deprivation of liberty may thus happen in a spatial continuum going from confinement in a cell to the free society. The distinctions between prison and free society and between prison-time and freedom-time are now blurred. As such, it may no longer be possible to speak about imprisonment, but about 'liberty regulation trajectories'.

In this new carceral space-time, prisoners are facing new demands. First, they must be active, constantly. This *activity* is not about mere functioning, but about every sort of activity, by every actor inside the prison, without distinction of function or status. This constant activity must be grounded in the *activation* of everyone, which means that every participant in the prison-project is expected to take initiatives, to develop original and personal forms of activities. The system is ruled from the ground level and not through a top-down approach. It is predicated on the social, professional, and psychological mobility of the subject. The relation to the collective level does not happen through incorporation in populations managed from above (biopower), but through *participation* in projects. Victim restoration, training, therapies, and the prison itself are nothing more than temporary projects proposed to the participatory good will of partners by the inmate, the victims, the administration, and so on.

In order to efficiently participate, *adaptation* is a key value. While prison is identified as a threat for one's autonomy through a total support (in the sense of a total institution) causing dependency, adaptation is valued as the condition to participate in as many projects as possible and to fully reintegrate into the (more) free society. The latter is therefore seen as a field where adaptability and initiative are key skills.

Legitimate prison in a mobilitarian world

What can be seen then, in the Belgium Prison Act, is a discourse about prison adapted to the contemporary ever-dominant ideological system. In other texts (Mincke, 2013, 2014a, 2016; Mincke and Lemonne, 2014), it was proposed that the idea of a shift in the social construction of space-time can be linked to a modi-fication of the social relation to mobility and to the growing praise of mobility for itself under the form of a mobilitarian ideology.

From limit-form to flow-form

The classical prison was based upon the representation that space can be given a sense through borders and that borders depended on temporal stability. The so-called limit-form spatiotemporal morphology is no help in understanding the new conception of prison. But we can link it with another model: the flow-form. This morphology is based on an ever-flowing time. Change is constant – even if it may be slow – and it prevents borders from gaining stability. From one day or year to another, situations change in such a way that the very idea of a stable organisation of space through well-defined boundaries cannot be conceived. It is useless trying to define borders, because time flows and constantly changes the reality, so that a border has to be constantly revised to reflect reality. The question is not that real-ity suddenly becomes unstable, but rather that the social construction of time has changed, resulting in a different way to deal with change.

This particular conception of time means that space cannot be structured based on borders. Does this mean that it remains unstructured? Certainly not. It is organised by attraction poles. The spatial localisation of an object is not linked to its position inside or outside certain boundaries, but to its relations with other objects. This is because every object, be it important or not, is an attraction pole. Localisation is thus a relationship, rather than a situation in a space structured by itself. This implies that the relation to space, instead of being exclusive – e.g., you are inside or outside the prison – is cumulative. You are both in the free world and serving a prison penalty, or you are inside prison, but able to circulate in a certain and variable way. There is no longer a clear distinction between pure incarceration and total freedom, but a varying relation to a carceral continuum. Certainly immobilisation was never absolute, but it was viewed through binary categories (inside/outside). It is now more complex, since the relation with a pole is not conceived as exclusive from relations with (many) other poles.

Mobility test

The flow-form space is a networked space, rather than a bordered one. In this context, the question of mobility gains a new sense in the contemporary prison. In the limit-form, mobility was purely a crossing mobility, based on traversing the border between two spaces. The movement implied a pre-existent embedding in a location, a planning, a determined meaning, and a fixed goal. But in a space without borders, there can arguably be no crossings. And if one's position is determined by its reciprocal relations with other entities, and if time constantly flows and modifies the context, there can be nothing other than movement. Mobility becomes irrepressible and constant. It makes no sense to describe imprisonment as a stay in a cell for a defined stretch of time if prison's space-time is conceived under the flow-form. Each relation to this space-time can be nothing more than a trajectory. The prison building is just one pole among many others that constitute the carceral continuum. The doors must remain as open as possible, as everyone – inmates, social workers, lawyers, visitors, workers, et al. – are just passers-by. Mobility is the rule, the prime experience; immobility can only be the result of a constraint, of an effort that cannot be maintained for a long period.

More than that, in a world of movement, immobility is a danger: the danger of a desynchronisation with the society. Or it can be a symptom, the symptom of the loss of flexibility. Immobility is the sign or the cause of a problematic relation with a moving world. In this context, mobility has to become compulsory. The mobility we are talking about is not only a physical one, but also social, moral, professional, or family mobility, since mobility is nothing but a displacement in space over a stretch of time. For a sociologist, space is the result of a process of spatialisation, applied to physical and non-physical realities (Mincke, 2014b; Kaufmann and Mincke, forthcoming). When the inmate is encouraged to come into contact with his victim and to propose a restorative process, the aim is to change their mutual positions and to help both of them to modify their trajectory, which was strongly disrupted by the offence.

As we can see, in this space-time morphology, immobility is problematic and mobility is praised for itself. This system of thought is what we call 'mobilitarian ideology'. It can be summarised as imposing four imperatives.

Activity demands a constant action, be it professional, personal, recreational, or lucrative. It implies constantly tracking the activity of people and systems and looking forward to a maximum efficiency. Not a second can be lost and there is no room for downtime. In this respect, prison-time cannot be a lost period anymore, but must be used at its maximum. The same is demanded, for instance, from retired persons who must be active senior citizens, from the unemployed who must search for a job, whether it exists or not, or from organisations, which are supposed to be constantly overworked.

Activation imposes that persons and organisations be at the principle of their own movements. They are not supposed to wait for commands or to strictly apply pre-existent norms, but to take initiatives, to imagine new activities, to have their own goals. Accordingly, inmates are asked to write a detention plan, explaining how they intend to use their time in prison. No one else but them knows what to make of themselves, there is no expert to say what the 'treatment' should be, and each person is required to make his own projects. In the same way, social beneficiaries are intended to take an active part in their social and professional reintegration, litigants are proposed to personally take their conflict in charge through mediation, or workers are demanded by the management to show initiative in work organisation.

The projects they have to conceive are collective ones. There is no question of meditation or spiritual ambitions. What is demanded is to develop participatory processes, which temporarily assemble partners in pursuit of a common goal. *Participation* is therefore a central imperative. Prison itself, is no longer seen as an institution that seizes individuals, but as a common project within which everyone has to be involved. In the same way, participatory pedagogy, democracy, management, or dispute resolution are seen as positive ways to renew the answer to old questions. Almost everything that used to be considered as structured institutions ruled by a hierarchical principle, from work to politics, and from education to family, is now seen as a succession of projects, associating free-willing partners.

In order to link up projects and be part of as many of them as possible, one has to be flexible. *Adaptation* is therefore a key imperative. The inmate has to adapt to the needs of the victim, to the offer of reintegrative support in his prison, to the professional perspectives after serving his time, etc. More generally, the inmate is asked to demonstrate flexibility, which, according to the preparatory documents, shows an ability to reintegrate society. Flexibility has become a motto: administration, companies, workers, litigants, parents, or politicians are expected to be able to change their views and actions in real time, in pursuit of a constantly changing context.

Through these four imperatives, mobilitarian ideology imposes mobility for people and organisations in every space they are committed to. This imperative can be traced in family life, management, political organisation, productive systems, and so on. In such an ideological context, the inmate is seen as someone who must not lose, and must even reinforce, an ability to freely move in the society, and painful immobility is seen as part of the problem of incarceration, rather than as a solution.

This is why the prison is no longer meant to cause suffering through immobility (painful immobilisation), but rather to test inmates about their mobility potential (mobility tests). This means that the prison is not presented as a closed (physical, social, legal, etc.) space within which one endures a special condition, but as an open space where people's trajectories cross each other. The aim is not to confine, but to reorient one's trajectory. The question is not if detainees have served their penalty, but if they can be afforded more freedom regarding the direction of their trajectory. Whether they prove to be able to respect mobilitarian imperatives will help to determine the liberty level they will enjoy. In this sense, the aim does not lie in prison itself; the prison is just a means to pursue other aims that will be pursued further on the outside.

Conclusion

In short, a new discourse about prison has emerged, which presents it as a place where the mobility-skills of inmates can be tested. If they show abilities, they can be authorised to continue their trajectory towards complete freedom. Incarceration is, from the beginning, seen in the context of release, but strict conditions are set to benefit from a lighter regime. In light of the Belgium Prison Act, planning, restorative procedures, and personal initiatives are, on one hand, means to reintegrate the inmate back into society, but also, on the other hand, provide challenges for them to prove their mobilitarian skills. Those who do not prove an ability to comply with the mobilitarian imperatives may well be maintained in strict detention until the end of their term. The opening of prison space and its shift from immobilisation to mobilisation is thus a relative and conditional one. As we can see, we now face a very new vision of prison.

In an ideological context of compulsory mobility, of flowing time and open space, it is no wonder that the legitimation of prison through immobilisation is not quite as efficient as it once was. But, as already happened in the past, the radical devaluation of the legitimising supports of prison do not imply that it is weakening. The prison has already shown its ability to change its justifications. This is what is now at work, in the preparatory documents of the Prison Act. We do not claim that these visions are applied or even applicable, merely that they propose a particular conception of what a *legitimate* prison could be and of the project the carceral administration should have. The next question will be the way these discourses will change the prison in itself and the way prison will be able to remain unchanged, even though this would be incompatible with its new guidelines. This will depend on its ability to produce a discourse of conformity, and on whether people wish to see the cracks in the painting or to believe in mobilitarian fables.

From painful immobilisation to mobility tests is a major shift. It also seems to be a paradoxical one, since prison is the symbol of immobilisation. Accordingly mobility and the mobilitarian ideology are incredibly powerful. A mobilitarian ideology can be used to justify the prison system only if it is a strong and versatile discourse, if its attractiveness seems strong enough to convince us of its efficacy to legitimate anything, including an immobility-based institution. But, have we not heard time and time again that locking people behind bars would help to reintegrate them? As such, is the notion of mobility less absurd?

Notes

1 Law on 'principles governing the administration of prison establishments and the legal position of detainees', the Act of 12 January 2005.
2 Loi du 27 novembre 1891 pour la répression du vagabondage et de la mendicité (Law for the repression of vagrancy and begging).
3 All the quotes are translated for the purpose of this text. Official versions are in French and Dutch.
4 This situation is parallel with management theories, which exclude the question of the aims. The problem is not that right things are done, but that things are rightly done (Kaminski, 2002: 89). Management theories induced many changes in criminal policies (Mincke, 2013; Kaminski, 2002, 2003, 2004, 2008).

References

Bauman Z (2000) *Liquid Modernity*. Cambridge, UK; Malden, MA: Polity Press Blackwell.
Cresswell T (2006) *On the Move: Mobility in the Modern Western World*. New York, NY: Routledge.
Decroly V and Van Parys T (2001) Rapport final de la commission « loi de principes concernant l'administration pénitentiaire et le statut juridique des détenus ». Rapport fait au nom de la commission de la Justice par Vincent Decroly et Tony Van Parys. *Documents parlementaires, Chambre* (50-1076/001).
Delwit P and Gotovitch J (1996) La peur des rouges. In: *La peur du rouge. Histoire, économie, société*. Bruxelles: Editions de l'Université de Bruxelles, vii–xv.
Demonchy C (2004) L'architecture des prisons modèles Françaises. In: Artières P and Lascoumes P (eds) *Gouverner, enfermer. La prison, un modèle indépassable?* Paris: Presses de Science Po, 269–293.
Dreyfus HL, Rabinow P and Foucault M (1984) *Michel Foucault, un parcours philosophique: au-delà de l'objectivité et de la subjectivité*. Paris: Gallimard.
Erdman F, Bourgeois G, Coveliers H, Dardenne M, Herzet J, Hove G, Lalieux K, Talhaoui F and Van Parys T (2003) Proposition de résolution relative au rapport final de la Commission « loi de principes concernant l'administration pénitentiaire et le statut juridique des détenus ». *Documents parlementaires, Chambre* (50-2317/001).
Kaminski D (2002) Trouble de la pénalité et ordre managérial. *Recherches sociologiques* (1): 87–107.
Kaminski D (2003) Le management de la prévention. In: *Prévention et politique de sécurité arc-en-ciel. Actes de la journée d'études du 28 mars 2003*. Réseau interuniversitaire sur la prévention, 57–72.
Kaminski D (2004) L'affiliation managériale de la pénalité. In: *Judiciaire et thérapeutique: Quelles articulations?, Actes du colloque tenu à Bruxelles le 5 décembre 2003*. Judiciaire et thérapeutique: quelles articulations? Bruxelles: La Charte, 45–56.
Kaminski D (2008) Pénalité, management, innovation. *Revue de droit pénal et de criminologie*: 867–886.
Kaufmann V and Mincke C (forthcoming) Pour une sociologie des espaces non physiques. Working paper. Available at: http://hdl.handle.net/2078.3/165744
Maes E (2009) Van gevangenisstraf naar vrijheidsstraf. *200 jaar Belgisch gevangeniswezen*. Antwerpen/Apeldoorn: Maklu.
Mincke C (2013) Mobilité et justice pénale. L'idéologie mobilitaire comme soubassement du managérialisme. *Droit et société* (84): 359–389.

Mincke C (2014a) La médiation pénale, contre-culture ou nouveau lieu-commun? Médiation et idéologie mobilitaire. In: Beckers C, Burssens D, Jonckheere A and Vauthier A (eds) *Médiation pénale. La diversité en débat. Bemiddeling in strafzaken. En wispelturig debat.* Antwerpen/Apeldoorn: Maklu, 85–110.

Mincke C (2014b) L'espace est-il une dimension physique? Sociologie de l'espace ou spatialisation de la sociologie? In: *Penser l'espace en sociologie.* Bruxelles: Dépôt institutionnel de l'Académie Louvain. Available at: http://hdl.handle.net/2078.3/144668

Mincke C (2016) From mobility to its ideology: When mobility becomes an imperative. In: Endres M, Manderscheid K and Mincke C (eds) *The Mobilities Paradigm: Discourses and Ideologies.* London: Ashgate, 11–23.

Mincke C and Lemonne A (2014) Prison and (im)mobility. What about Foucault? *Mobilities* 9(4): 528–549.

Montulet B (2005) Au-delà de la mobilité: des formes de mobilités. *Cahiers internationaux de sociologie* 118: 137–159.

Afterword

Ordinary mobilities, ordinary incarcerations

Peter Merriman

> A travelling incarceration. Immobile inside the train, seeing immobile things slip by. What is happening? Nothing is moving inside or outside the train. ... A bubble of panoptic and classifying power, a module of imprisonment that makes possible the production of an order, a closed and autonomous insularity – that is what can traverse space and make itself independent of local roots. (de Certeau, 1984: 111)

In his short chapter on 'Railway incarceration and navigation' in *The Practice of Everyday Life*, Michel de Certeau reflects upon the 'incarcerational and naviga-tional' qualities of railway travel, in which the ordered movements and embodied actions of the passenger are constrained by the spatial confines and engineered environments of the railway carriage (de Certeau, 1984: 113). When considered in the context of the case studies outlined in *Carceral Mobilities*, it would be tempting to dismiss these as rather trivial remarks on the privileged movements and confinements of the railway passenger. The average fare-paying railway pas-senger experiences freedoms, choices, and comforts that are far removed from the experiences of those whose mobilities are enforced or severely constrained, whether migrants trafficked in lorries, refugees crammed on to boats, or prisoners transported in trains (Gigliotti, 2009; Presner, 2007). And yet, one can observe parallels between the ordinary and everyday incarcerations experienced by pas-sengers in public transport systems and discourses surrounding the ordering and incarceration of dangerous, deviant, and vulnerable bodies. Discourses of security, safety, efficiency, order, and segregation pervade discussions of the movements of both willing and unwilling, and voluntary and forced passengers. Passengers of all kinds may feel a sense of disempowerment, helplessness, frustration, and lack of control, as their movements are shaped and governed by state powers, expert authorities, and those seeking to enable and channel movements for political and economic gain. For some, this surrendering of freedom and control may be a com-fort, with the delegation of effort and responsibility to others allowing them to focus their attention on other matters. For others, such as the art critic John Ruskin writing in the nineteenth century, this incarceration and disengagement from the landscape transformed passengers into a piece of luggage, rendering them 'a mere parcel' (Schivelbusch, 1980: 59).

Sociologists and historians of mobility and technology have, at various times, remarked upon the gradual tendency for designers to cocoon the mobile subject inside the vehicle, insulating them from the environments and atmospheres through which they travel (see Bijsterveld et al., 2014; cf. Bull, 2004; Merriman, 2007; Urry, 2000). These changing practices of everyday 'insulation' (and, by implication, changing modes of engagement) can be closely aligned with changing practices of vehicle design and inhabitation, from the 'cocooning' of passengers inside air-conditioned cars, trains, and aeroplanes, to the adoption of automatic door-locking systems on these same vehicles. As I pull away in my modern hatchback car, the doors lock automatically when I reach a certain speed. My local trains have automatic door-closing and locking mechanisms controlled by the guard or driver (who deactivates the automatic door buttons on departure). As a rail passenger, I am 'incarcerated' in the train until the next station stop – subjected to railway 'bylaws', company policies, the actions of the driver, and the etiquettes of railway travel. There is, of course, nothing remarkable about these practices of everyday or ordinary incarceration, which are common throughout complex Western transport systems and, I assume, are for my own safety and security. This is a state of incarceration to which I willingly submit, retaining freedom to move around and travel in relative comfort.

Such examples of privileged incarceration can help us to address some of the initial questions posed by Peters and Turner in the introduction to this volume, namely: 'what is the carceral?' and what are the 'everyday spaces… where carceral conditions and qualities emerge through im/mobility practices'? While it would be all too easy to focus entirely on the carceral spaces and mobilities associated with the prison or detention centre, there are many different ways in which the movements of human bodies, non-humans, and other things are constrained at different spatial scales, whether in securing borders and boundaries, controlling populations, protecting private property, securing the flows of commodities and capital, or ordering animate non-human bodies. The carceral geographies and mobilities of a broad range of spaces, practices, subjects, and things could provide an important focus for future study, whether zoos, gated communities, financial systems, or entire nation-states and national territories.

References

Bijsterveld K, Cleophas E, Krebs S and Mom G (2014) (eds) *Sound and Safe: A History of Listening Behind the Wheel*. Oxford: Oxford University Press.

Bull M (2004) Automobility and the power of sound. *Theory, Culture and Society* 21(4–5): 243–259.

de Certeau M (1984) *The Practice of Everyday Life*. London: University of California Press.

Gigliotti S (2009) *The Train Journey: Transit, Captivity and Witnessing in the Holocaust*. Oxford: Berghahn.

Merriman P (2007) *Driving Spaces: A Cultural-Historical Geography of England's M1 Motorway*. Oxford: Blackwell.

Presner T (2007) *Mobile Modernity: Germany, Jews, Trains*. New York, NY: Columbia University Press.

Schivelbusch W (1980) *The Railway Journey: Trains and Travel in the 19th Century*. Oxford: Basil Blackwell.

Urry J (2000) *Sociology Beyond Societies*. Abingdon: Routledge.

Index

Printed and bound by CPI Group (UK) Ltd, Croydon, CR0 4YY

21/10/2024

01777055-0012